STRESS: A User's Manual

A Problem-Oriented Computer Language for Structural Engineering

STRESS: A User's Manual

A Problem-Oriented Computer Language for Structural Engineering

The Department of Civil Engineering
Massachusetts Institute of Technology

Steven J. Fenves
Robert D. Logcher
Samuel P. Mauch
Kenneth F. Reinschmidt

The M. I. T. Press
Massachusetts Institute of Technology
Cambridge, Massachusetts

Second Printing, September 1964
Third Printing, December 1966
Fourth Printing, August 1967

Library of Congress Catalog Card Number: 64-19174

Printed in the United States of America

PREFACE

The purpose of STRESS is to facilitate the use of digital computers for the analysis of structures. In order to accomplish this purpose the communication between the structural engineer, who is normally not a programming expert, and the computer must be improved. The STRESS system is an attempt to achieve this objective by use of a problem-oriented input language that enables the engineer to write the complete input program for the solution of a structural problem even though he has had no programming experience.

The purpose of STRESS is not to make possible the solution of unusually large or complex structural problems, although its capability in this regard is considerable. The principal virtue of the system is that it can handle a wide variety of structural analyses with a minimum of programming effort. For example, STRESS can analyze structures in two or three dimensions, with either pinned or rigid joints, with prismatic or nonprismatic members, and subjected to concentrated or distributed loads, support motions, or temperature effects. With this capability available, we hope that the practicing engineer will find it economical to use a computer on a day-to-day basis for the solution of routine structural problems. In the educational field, STRESS will make possible the routine use of computers in the teaching of structural design.

Another important aspect of the system is the ease with which the engineer may obtain additional solutions for modifications of the original structure. This is particularly important in design that involves the evolution of a satisfactory structure by a trial process.

The STRESS version presented here is not considered to be a final or complete product. Work is currently under way to improve, refine, and add capabilities to the system; and it is expected that this will continue indefinitely.

This User's Manual is an attempt to present as concisely as possible the STRESS input language. The information contained herein should enable the engineer to fully utilize the present system. The internal programming is described in detail in the forthcoming STRESS Reference Manual, which is intended to enable extensive users to modify the system to suit best their own purposes.

Work on STRESS began in the Fall of 1962 under the direction of Professor S. J. Fenves of the University of Illinois who was a visiting member of the M.I.T. faculty during the year 1962-1963. The project staff included Professor R. D. Logcher, Professor S. P. Mauch, Mr. K. F. Reinschmidt, and Mr. R. L. Wang. In the Fall of 1963 the project was placed under the general supervision of Professor J. M. Biggs with Professors Logcher and Mauch directly in charge of the programming effort. During the ensuing period important contributions were made to the debugging of the system by Professor Z. M. Elias, Mr. R. V. Goodman, Miss S. C. Finkelstein, Mr. S. G. Mazzotta, Mr. J. R. Roy, and Mr. C. Singhal.

The STRESS project has been partially supported from a major grant for the improvement of engineering education made to M.I.T. by the For Foundation. Additional support was provided by Project MAC, an M.I.T research program, sponsored by the Advanced Research Project Agency, Department of Defense, under Office of Naval Research, Contract No. Nonr-4102(01). Reproduction in whole or in part is permitted for any purpose of the U.S. Government. The work was done in part at the M.I.T. Computation Center, and the aid and support of the Center and its personnel are gratefully acknowledged.

CONTENTS

INTRODUCTION

STRESS, which is the abbreviation for STRuctural Engineering Systems Solver, is a programming system for the solution of structural engineering problems on digital computers.

STRESS consists of (1) a language that describes the problem and (2) a processor (computer program) that accepts this language and produces the requested results. This manual explains the use of STRESS and is accordingly concerned primarily with the STRESS language.

A problem is described with the STRESS language by writing a number of statements specifying the nature and size of the structure, the loads, a solution procedure, and the results desired. Modification of any of this information may be requested to obtain additional results for slightly altered problems.

From the user's standpoint, STRESS has the following principal characteristics:

1. Communication with the computer is entirely in the engineer's language. A problem is described to STRESS in essentially the same terms as those which one engineer would use in instructing another engineer. Thus, the STRESS language is easily understood, and no conventional computer programming experience is needed to program in STRESS.
2. The description of a problem in STRESS consists both of the data for the particular problem and the procedures to be used in solving that problem. Thus, a new and unique program is written for each application, allowing the user complete freedom in specifying his problem within the scope of the STRESS language.
3. After a problem has been described to STRESS, modification of portions of the problem can be easily and conveniently specified, and the corresponding results produced. This facility makes STRESS a powerful structural design tool.

In comparison with standard structural analysis programs, STRESS provides far greater ease of use and flexibility than any program known to the authors. There are no rigid input forms, difficult-to-remember numeric codes, or arbitrary size restrictions. STRESS is capable of handling large structural problems but can handle small and medium size problems with equal efficiency.

The STRESS processor is designed to be flexible and easy to change or expand. A complete description of the processor and all information for changing the processor is given in STRESS Reference Manual.[1]

The present manual describes the current capabilities of the STRESS system. At present STRESS can perform the linear analysis of elastic, statically loaded structures composed of slender members. The present system represents only the initial development, and additional capabilities can and will be included in the future. Any structure that can be described with the language presented in Chapter 2 can be analyzed. The solution provides such information as member forces at the member ends, joint displacements, reactions, and so forth.

It is assumed that STRESS will be used by persons familiar with standard structural analysis. Familiarity with recent developments in structural analysis, such as matrix and network formulations, while desirable, is by no means required. However, Hall and Woodhead's Frame Analysis[2] is recommended for readers who desire a general introduction to the matrix analysis of structures. The stiffness method of solution is described briefly in Appendix 1 for the convenience of the reader.

This manual consists of four chapters. Chapter 1 is a primer, introducing the major components of the language and giving a sample problem. Chapter 2 describes in detail the statements available in the current STRESS language, their meaning, variation, and use. Chapter 3 contains rules for using STRESS on an IBM 709/7090/7094 computer, with particular emphasis on the procedure for the IBM 7094 at the M. I. T. Computation Center. Chapter 4 gives a brief description of the operation of the STRESS processor. Appendix B contains a description of the IBM 1620 language and use.

---

[1] S. J. Fenves, R. D. Logcher, S. P. Mauch, STRESS Reference Manual, The M. I. T. Press, Cambridge, Mass., forthcoming, 1964.

[2] A. S. Hall and R. W. Woodhead, Frame Analysis, John Wiley & Sons, New York, 1961.

Chapter 1

PRIMER

1.1 Scope of STRESS

The STRESS system is presently implemented to perform the linear analysis of elastic, statically loaded, framed structures. Analysis means the computation of joint displacements, member distortions, member end forces and reactions for a structure, given the makeup and orientation of all the members, and the type, position, and magnitude of all the applied loads, displacements, and distortions. The term "framed," or "lumped parameter," structure is used to denote structures composed of slender elements, that is, members that can be represented by their centroidal axis and analyzed as line elements. The structure itself may extend in two or three dimensions, and at any joint the members may be pinned or rigidly connected.

At present STRESS uses only the stiffness (displacement) method of analysis. This restriction, however, has no effect on the user who desires to obtain results for a particular structure.

Any structure that can be described by the language can be analyzed. Some computations may have to be performed outside of the STRESS system to analyze a structure that cannot be completely described by the language. For example, the load-deflection properties of a curved member could be supplied as either the member stiffness or flexibility matrices since STRESS cannot as yet determine these properties. Similarly, a shear wall could be approximated by a lattice analogy and the lattice members used to describe the structure. In most cases, however, no external computations are necessary.

1.2 Coordinate Systems

The analysis of framed structures deals with forces (force resultants, actions) and displacements or distortions. In STRESS, all components of force and displacement vectors are described in right-handed, orthogonal Cartesian coordinate systems. Such a coordinate system is shown in Figure 1.1, where x, y, and z denote coordinate directions and u_1

through u_6 denote the six components of a force or displacement vector.

In dealing with the description of a problem (input data) and the results produced by STRESS (output data), it is necessary to distinguish between global and local coordinate systems.

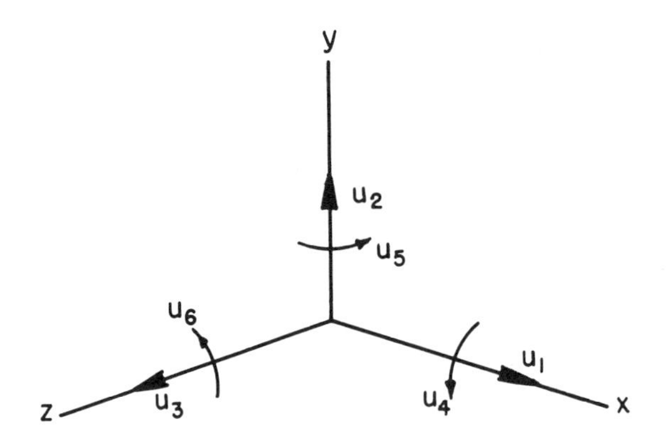

Figure 1.1. Coordinate system.

The global coordinate system is an arbitrary system, usually chosen so that the direction of the axes coincide with the major dimensions of the structure. All joint data are specified in terms of the global system; and the computed joint displacements and reactions are similarly output in the same system. The joint coordinates are specified with respect to an arbitrary origin, which may (but need not) be chosen at one of the support points.

Figure 1.2 illustrates the use of global coordinates for a space and a plane structure. A plane structure must be located in the X-Y plane of the three-dimensional coordinate system.

A local coordinate system is associated with each member, and all member data are specified in terms of this system. The local x axis coincides with the axis of the member, and its direction is from the start of the member to its end where "START" and "END" are specified in the input data. (See Section 2.4.3, MEMBER INCIDENCES). The y and z axes coincide with the principal axes of the member, as shown in Figure 1.3. For plane structures, it is assumed that one of the principal axes of the member lies in the global X-Y plane of the structure. (This is required in order to avoid out-of-plane deformations in a plane frame or in-plane deformations in a plane grid.)

1.3 Relationship Between Global and Local Coordinates

The position of a member in space is determined by the global coordinates of its end points. However, unless the member is axially symmetric, there is one unspecified degree of freedom, that is, the rotation

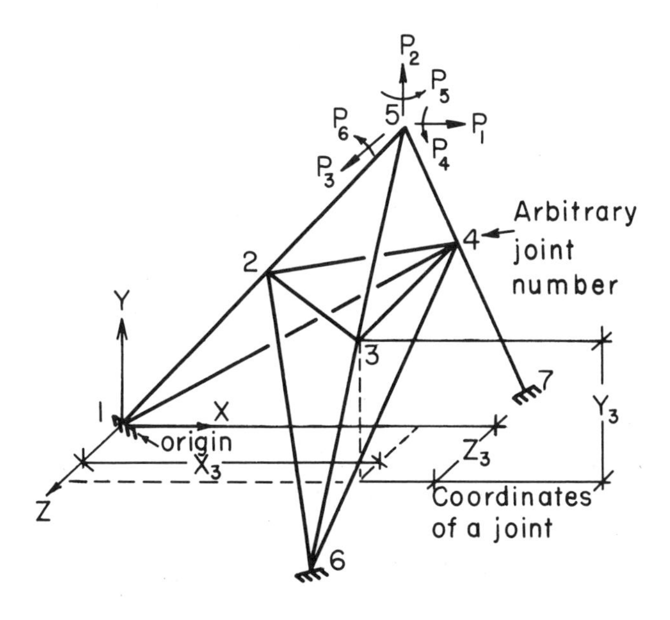

a. Space truss

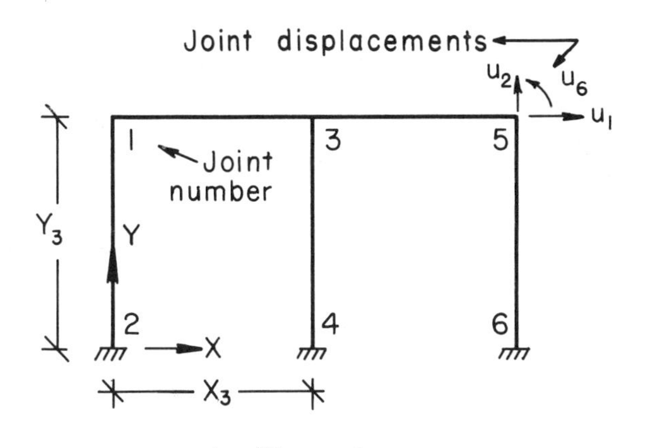

b. Plane frame

Figure 1.2. Use of global coordinates.

of the principal axes of the member from the global axes. This additional quantity is called the angle β, BETA (see Figure 1.3.)

For differentiation, call X, Y, Z a global coordinate system and x, y, z a member system. Let A be a plane containing the member x axis and a line parallel to the global Y axis (therefore perpendicular to the X-Z plane). Let y' be a coordinate in this plane and perpendicular to the x axis. The direction of y' must be taken so that the projection of y' on the Y axis is in the positive Y direction. Then β is the angle from y' to y, positive by the right-hand rule around x. This definition is not sufficient if the x axis

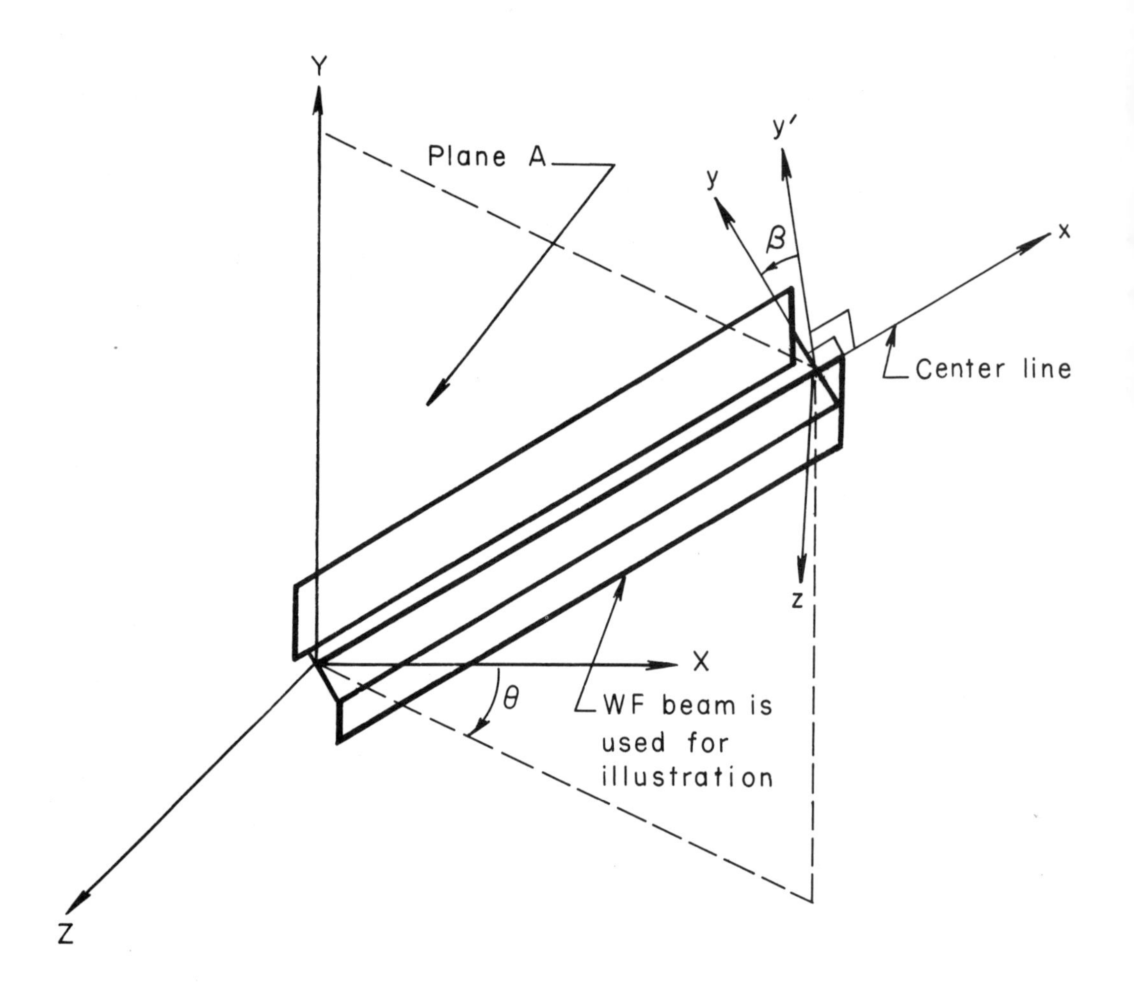

Figure 1.3. Local coordinates for a member.

is parallel to the Y axis, in which case the plane A is indeterminate. Then β is the angle from the $-X$ axis to the y axis if the x axis is in the same direction as the Y axis and from the $+X$ axis if not.

For plane structures, the definition is taken that when $\beta = 0$, the member z axis is parallel to and in the same direction as the Z axis. For plane structures, β must be either zero or a multiple of $90°$; β is given in decimal degrees.

1.4 Structural Types

STRESS allows considerable flexibility and simplification in the analysis of plane frames or grids and plane or space trusses. This is done by omitting the components of the general six-dimensional force or displacement vector that do not enter into the analysis. The number of degrees of freedom (that is, the size of the vectors) thus varies from 2 for plane trusses to 6 for space frames.

The vector components used throughout the system are given in Table 1.1 for the available structural types. In the table, u_1, u_2, u_3 denote forces or displacements in a right-handed coordinate system, and u_4, u_5, u_6 denote moments or rotations about the axes as shown in Figure 1.1. These components pertain to both member and joint quantities, except for truss members. For plane and space trusses, the vector components shown are those for global coordinates. For local coordinates, obviously there is only one component, u_1, the axial force or distortion of the member. For any structural type, it is necessary to specify only those input quantities (loads, displacements, member section properties, and so forth) corresponding to the vector components listed for the particular type.

Table 1.1. Vector Components for Structural Types

Type of Structure	Degrees of Freedom (JF)	Components Used
Plane Truss	2	u_1 u_2
Plane Frame	3	u_1 u_2 u_6
Plane Grid	3	u_3 u_4 u_5
Space Truss	3	u_1 u_2 u_3
Space Frame	6	u_1 u_2 u_3 u_4 u_5 u_6

For plane frames, the use of local components u_2 and u_6 implies that the local y axis is in the plane of the structure. A moment on a member is then about the z axis, and the quantities A_X (cross-sectional area), A_Y (shear area), and I_Z (moment of inertia about z axis) are needed to define the properties of a cross section. It is possible to specify $\beta = 90°$ and reverse these last subscripts. The plane grid, with the same axis arrangement ($\beta = 90°$), requires A_Y, I_Z, and I_X (torsional constant). For plane structures, a β angle is needed and can be used only if steel section names are used.

1.5 The Use of Releases

STRESS assumes that all force or displacement vector components of a member or joint are related by identical continuity and equilibrium equations. For example, a support joint in a space frame provides force and moment reactions in all three component directions, or the equilibrium equation at a joint in a plane frame involves displacement components in the X and Y directions as well as rotation about the Z axis.

In many structures, local deviations from this pattern may occur. The introduction of hinges, rollers, and so forth, makes certain force components equal to zero.

In order to avoid ambiguities in designations such as hinges or rollers, in the STRESS language the word RELEASE is used to specify zero force components.

Two types of releases are implemented in STRESS.

Joint releases may be specified at support joints to indicate zero reaction components, that is, joint displacement components which are not prescribed. Released components need not be in global coordinate directions, but must be orthogonal at a joint.

The joint release orientation is shown in Figure 1.4 by the x', y', z' coordinate system. The rotation of this system from the global coordinate system is given by the angles θ_1, θ_2, θ_3 as follows:

1. θ_1 is the angle from the X axis to the projection of the x' axis on the X-Y plane.
2. θ_2 is the angle from the projection of the x' axis on the X-Y plane to the x' axis. The positive direction is measured from the X-Y plane toward the Z axis.
3. θ_3 is measured from a plane including the x' and Z axes, from the projection of the z' axis on this plane to the z' axis. The positive direction is measured about the x' axis by the right-hand rule.

A joint release does not concern the fixity of the members at that joint, only the fixity of the joint to the support, as shown in Figure 1.5.

Member releases may be specified only at the member ends. A member release indicates that a force component is zero. Released components must be in the local member coordinate system. Member release information at the member end must be given if full fixity of the member connection is not desired. In the case when full fixity does not exist at a support joint incident to only one member, member release at that end has the same meaning as a joint release. One restriction on this alternate specification exists, and it arises from checking load data for consistency. Joint loads can be given only at joints that are free to displace. A support joint that has certain released components has such freedom. Therefore, joint releases must be specified for any support joints carrying applied loads.

For structures with members and joints corresponding to more than one type, such as a truss with certain joints rigidly connected, STRESS requires that the type with the greater number of components be specified and that local deviations from this type be specified separately. Thus, if the structure contains one or more rigid joints, it would be classified as a frame and the hinges introduced by means of RELEASE statements.

1.6 Units

At present, STRESS performs no conversions of units. Thus, all lengths and forces must be input in consistent units. For example, dimensions, applied displacements, and other length data may be in inches,

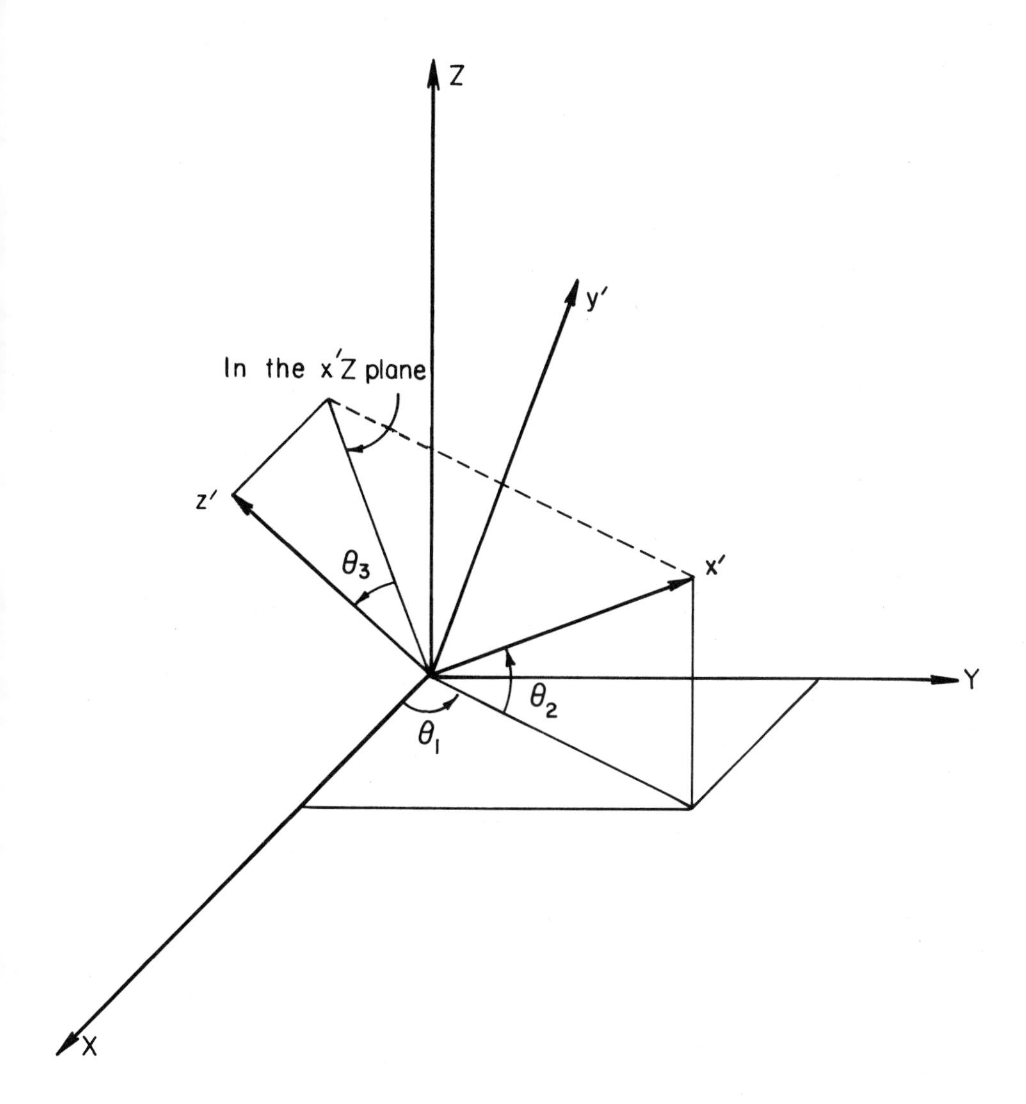

Figure 1.4. Specification of joint release orientation.

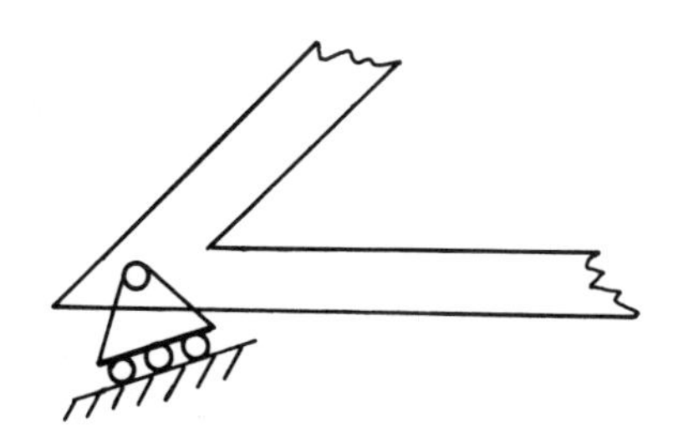

Figure 1.5. Member fixity with joint releases.

concentrated forces in kips, moments in inch-kips, distributed loads in kips/inch, and so forth. The computed results will be in the corresponding units.

The modulus of elasticity E is taken as 1.0, and the shearing modulus G as 0.4, if not given, but may be specified by the CONSTANTS Statemen (see Section 2.4.7). Therefore, if E is not given, input displacements ar distortions must be multiplied by E, and displacement and distortion results divided by E.

1.7 Identification of Joints, Members, and Data

In STRESS, joints and members are referred to by identification numbers assigned by the user. Numbering and ordering of data for members and joints are arbitrary with one restriction: for an initial problem (that is, not a modification), member and joint identification numbers may not exceed the specified number of members and joints, respectively. Any member or joint identification number may be used with modifications, but, for efficiency, it is best to keep these numbers as low as possible.

For member and joint information, a tabular form or input is generall used. A heading statement described in Sections 2.4 and 2.5 initiates th tabular input mode. When this mode is used, a member or joint number must be the first item in subsequent statements. The tabular mode is te minated when the first item is not an integer number. The data associate with the various statements (such as joint coordinates, member properti are generally identified by appropriate labels. Two deviations are permitted. For the various types of member and joint information, a descriptive rather than a tabular form may be used. For the data, labels may be omitted provided that a fixed order is maintained. These alterna forms are described in Section 2.7.

1.8 Statement Types

Input to the STRESS system consists of problem-oriented statements using common engineering terminology. The usual distinction between process descriptors (FORTRAN source language) and data has been com pletely eliminated, and the majority of statements contain both kinds of information.

The following classification is intended primarily to indicate the scope of the system. The exact function of each statement is described in deta in Chapter 2.

Header statement. The word STRUCTURE followed by any identifying information serves to start a new problem.

Size descriptors. Several statements are needed to define the size of th problem to be handled. These include:

NUMBER OF JOINTS

NUMBER OF SUPPORTS
NUMBER OF MEMBERS
NUMBER OF LOADINGS

Process descriptors. Statements in this category give information about the procedures to be used for a particular problem. These statements include:

TYPE
METHOD
TABULATE
SELECTIVE OUTPUT
PRINT

Structural data descriptors. To describe completely a framed structure, it is necessary to provide information about its geometry, topology (interconnection of members and joints), mechanical properties (load-deflection relationships of the members), and the presence of local releases (such as hinges or rollers). Six types of statements are provided:

1. Geometry is specified in terms of joint coordinates by the statement

   JOINT COORDINATES

   followed by the X, Y, Z coordinates of each joint (or X, Y for plane structures). These statements are also used to describe the status (that is, free or support) of the joints.
2. The presence of hinges or rollers at support joints is given as

   JOINT RELEASES

   followed by the joint numbers and the designation and orientation of the released (zero) force components.
3. The interconnection of the members is specified by the statement

   MEMBER INCIDENCES

   followed by a list giving the starting and ending joint of each member. The meaning of this statement is best illustrated by the descriptive input form, which for a typical member may be MEMBER 17 GOES FROM JOINT 10 TO JOINT 7.
4. The load-deflection properties of the members are specified as

   MEMBER PROPERTIES

   followed by a statement for each member giving the type of member, and the labels and numerical values of the properties.
5. The presence of hinges in the members is given as

   MEMBER RELEASES

   followed by the member numbers and the position and orientation of the released force components.
6. Constants associated with the members are specified by the

CONSTANTS

statement.

Loading data descriptors. The loading applied to the structure is specified in terms of loading condition descriptors, descriptors of individual loads, and descriptors of groups of loads, as follows:

1. The word

   LOADING

   followed by any identifying information delineates groups of loads, together comprising a loading condition, and serves as a loading condition header.
2. Individual loads are specified by statements such as

   JOINT LOADS

   followed by the joint numbers and the components of applied load,

   JOINT DISPLACEMENTS,

   MEMBER DISTORTIONS, and

   MEMBER LOADS

   followed by a statement for each load giving the member number, the orientation, magnitude, and type of the load.
3. Certain loading specifications involve general information such as

   COMBINE

   followed by a list of loading conditions to be combined.

Modification descriptors. To permit rapid evaluation of alternate designs the following statements can be used after an initial problem has been defined:

MODIFICATION	with information for output identification
ADDITIONS CHANGES DELETIONS	interspersed with pertinent statements of all the above types describing the modification

Termination statements. These statements terminate the input of portion or all of the statements of a problem. They are

SOLVE
SOLVE THIS PART
FINISH

1.9 A Sample Problem - Initial Specification

As an introduction to the scope and capability of STRESS, the simple plane frame shown in Figure 1.6 will be analyzed for two loading conditions: (1) uniform load of 1.2 kip/ft (= 0.1 kip/inch) on all horizontal members, and (2) horizontal loads of 20 kips on the two floors, as shown in the figure. The entire STRESS program is developed in this section, and Section 1.10 shows how a problem may be modified. It is suggested that the reader return to this section after he has studied the specific description of the statements in Chapter 2.

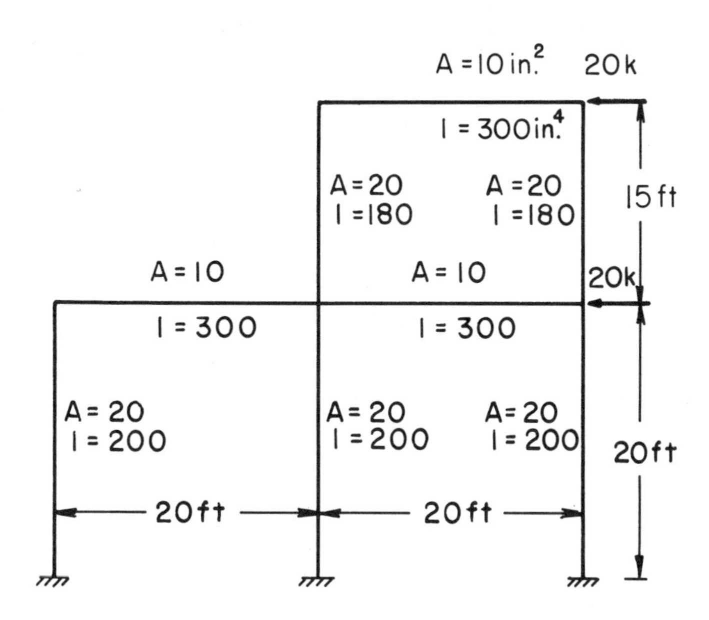

Figure 1.6. Sample structure.

The first step is to number the joints and members as shown in Figure 1.7. The directions shown for the members are arbitrary and chosen for convenience. Similarly, the origin of global coordinates is arbitrarily chosen at joint 5. Consistent units are chosen as inches for length and kips for force, requiring the dimension changes illustrated in Figure 1.7. The description of the problem follows closely the order of statement types given in Section 1.8, although such an order is not mandatory. The header and size descriptors can be written down immediately:

```
STRUCTURE  SAMPLE STRUCTURE
NUMBER OF JOINTS      8
NUMBER OF SUPPORTS  3
NUMBER OF MEMBERS   8
NUMBER OF LOADINGS  2
```

The structure is a plane frame and is to be analyzed by the stiffness method. The next two statements are therefore

```
TYPE PLANE FRAME
METHOD STIFFNESS
```

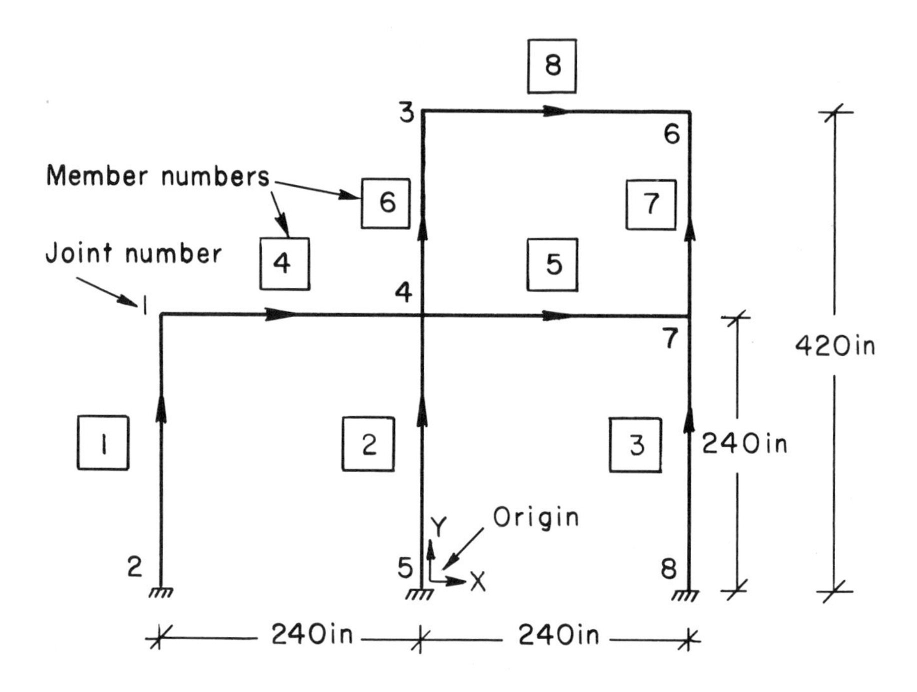

Figure 1.7. Numbering of members and joints.

Let us assume that member forces and reactions are desired for both loading cases. We therefore insert the statement

TABULATE FORCES, REACTIONS

at this point, before any individual loading has been identified. Additional output requests will be handled later.

We are now ready to describe the geometry of the structure by specifying the coordinates of all joints. We write the tabular header statement

JOINT COORDINATES

and follow by giving for each joint its number, X and Y coordinates, and status label (FREE, F, or blank for free joints, SUPPORT or S for fixed joints):

```
1 X -240. Y 240. FREE
2 X -240. Y   0. SUPPORT
5 X    0. Y   0. S
8 X  240. Y   0. S
4 X    0. Y 240.
7 X  240. Y 240.
3 X    0. Y 420.
6 X  240. Y 420.
```

Note that the order of the joints is immaterial. Since there are no joint releases (all supports are fixed), we can proceed to the description of member incidences. We again use a header statement, followed by tabular input in the order: member number, joint number as start of member, joint number at end of member.

```
MEMBER INCIDENCES
1 2 1
2 5 4
3 8 7
4 1 4
5 4 7
6 4 3
7 7 6
8 3 6
```

The member properties can now be specified. We assume that all members are prismatic. Furthermore, since this is a plane frame (and in subsequent modifications we do not intend to analyze it as part of a larger space frame), we need only to specify the cross-sectional area A_x and moment of inertia I_z of the members. Following the appropriate header statement

```
MEMBER PROPERTIES
```

we write for a typical member

```
8 PRISMATIC AX 10.0 IZ 300.0
```

Since we have not labeled and given AY, A_y is set to zero, indicating that shearing deformations are to be neglected. The same result can be achieved for the next member by writing

```
4 PRISMATIC AX 10.0 AY 0.0 IZ 300.0
```

For the columns, we may use a new header, putting in the header the designation common to all the members:

```
MEMBER PROPERTIES, PRISMATIC
1 AX 20.    IZ 200.
2 AX 20.    IZ 200.
3 AX 20.    IZ 200.
5 AX 10.    IZ 300.
6  IZ 180. AX  20.
7  IZ 180. AX  20.
```

Obviously, all members could have been included in the second tabular form.

If we assume the structure to be steel, the modulus of elasticity can be specified for all members as

```
CONSTANTS E, 30000., ALL
```

This completes the description of the structure. The loading data for the first loading are specified as

```
LOADING 1 UNIFORM LOAD ALL BEAMS
MEMBER LOADS
8 FORCE Y, UNIFORM, -0.1
4 FORCE Y, UNIFORM, -0.1
5 FORCE Y, UNIFORM, -0.1
```

Note that the distributed load acts in the member y direction and that, with the member orientations shown in Figure 1.7, a downward load acts in the negative y direction. Also, omitting the distances to the start and end of the load means that the load extends over the entire member.

The second loading condition can be similarly described:

```
LOADING 2 WIND FROM RIGHT
JOINT LOADS
6 FORCE X -20.0
7 FORCE X -20.0
```

For this loading only, suppose we are also interested in joint displacements, so we add

```
TABULATE DISPLACEMENTS
```

This concludes the description of the problem, so we write

```
SOLVE THIS PART
```

If we want selective output in addition to the quantities requested, we might add, for example, the following statements:

```
SELECTIVE OUTPUT
LOADING 1
PRINT DISPLACEMENTS 6, 7, DISTORTIONS 8
LOADING 2
PRINT DISTORTIONS 8
```

The problem description is now complete except for the

```
FINISH
```

statement. Figure 1.8 shows the complete STRESS program. The results obtained are shown in Figure 1.9. Figure 1.10 shows both the selective output requests and results.

1.10 Sample Problem - Modifications

In order to illustrate the facility with which modifications can be performed in STRESS, assume that the structure analyzed previously is also to be analyzed with the following changes:

a. The second floor lowered to 30 feet above ground.
b. Column 1 battered on a 4 to 1 slope and hinged at the bottom.

```
STRUCTURE SAMPLE STRUCTURE
NUMBER OF JOINTS 8
NUMBER OF SUPPORTS 3
NUMBER OF MEMBERS 8
NUMBER OF LOADINGS 2
TYPE PLANE FRAME
METHOD STIFFNESS
TABULATE FORCES, REACTIONS
JOINT COORDINATES
1 X -240. Y 240. FREE
2 X -240. Y   0. SUPPORT
5 X    0. Y   0. S
8 X  240. Y   0. S
4 X    0. Y 240.
7 X 240. Y 240.
3 X    0. Y 420.
6 X 240. Y 420.
MEMBER INCIDENCES
1 2 1
2 5 4
3 8 7
4 1 4
5 4 7
6 4 3
7 7 6
8 3 6
MEMBER PROPERTIES
8 PRISMATIC AX 10.0 IZ 300.0
4 PRISMATIC AX 10.0 AY 0.0 IZ 300.0
MEMBER PROPERTIES, PRISMATIC
1 AX 20. IZ 200.
2 AX 20. IZ 200.
3 AX 20. IZ 200.
5 AX 10. IZ 300.
6 IZ 180. AX 20.
7 IZ 180. AX 20.
CONSTANTS E, 30000., ALL
LOADING 1 UNIFORM ALL BEAMS
MEMBER LOADS
8 FORCE Y UNIFORM, -0.1
4 FORCE Y UNIFORM, -0.1
5 FORCE Y UNIFORM, -0.1
LOADING 2 WIND FROM RIGHT
JOINT LOADS
6 FORCE X -20.
7 FORCE X -20.
TABULATE DISPLACEMENTS
SOLVE THIS PART
```

Figure 1.8. Initial problem specification.

Figure 1.9. Initial problem results.

STRUCTURE SAMPLE STRUCTURE

LOADING 1 UNIFORM ALL BEAMS
MEMBER FORCES

MEMBER	JOINT	AXIAL FORCE	SHEAR FORCE	BENDING MOMENT
1	2	10.5446852	-1.2292451	-92.6043816
1	1	-10.5446852	1.2292451	-202.4144402
2	5	38.9818921	0.4814302	44.0447655
2	4	-38.9818921	-0.4814302	71.4984941
3	8	22.4734185	0.7478165	65.6633883
3	7	-22.4734185	-0.7478165	113.8125582
4	1	1.2292448	10.5446855	202.4144363
4	4	-1.2292448	13.4553142	-551.6899033
5	4	-1.8463720	13.3659654	628.3942719
5	7	1.8463720	10.6340343	-300.5625610
6	4	12.1606119	-2.5941878	-148.2028809
6	3	-12.1606119	2.5941878	-318.7509308
7	7	11.8393848	2.5941872	186.7499962
7	6	-11.8393848	-2.5941872	280.2036972
8	3	2.5941897	12.1606133	318.7509384
8	6	-2.5941897	11.8393863	-280.2037277

LOADING 1 UNIFORM ALL BEAMS

JOINT	X FORCE	Y FORCE	BENDING MOMENT
		SUPPORT REACTIONS	
2	1.2292451	10.5446852	-92.6043816
5	-0.4814302	38.9818921	44.0447655
8	-0.7478165	22.4734185	65.6633883
		APPLIED JOINT LOADS	
1	-0.0000003	0.0000002	-0.0000038
3	0.0000019	0.0000017	0.0000076
4	0.0000013	-0.0000010	-0.0000153
6	-0.0000025	0.0000018	-0.0000305
7	0.0000013	0.0000004	-0.0000057

LOADING 2 WIND FROM RIGHT
MEMBER FORCES

MEMBER	JOINT	AXIAL FORCE	SHEAR FORCE	BENDING MOMENT
1	2	11.1952969	-13.3343194	-1776.2674866
1	1	-11.1952969	13.3343194	-1423.9691467
2	5	10.3773841	-14.7321048	-1890.3127136
2	4	-10.3773841	14.7321048	-1645.3924255
3	8	-21.5726805	-11.9339367	-1669.2037506
3	7	21.5726805	11.9339367	-1194.9410553
4	1	13.3343413	11.1952968	1423.9691315
4	4	-13.3343413	-11.1952968	1262.9020844
5	4	16.4668262	13.3848399	1434.1739960
5	7	-16.4668262	-13.3848399	1778.1875763
6	4	8.1878413	-11.5994368	-1051.6837006
6	3	-8.1878413	11.5994368	-1036.2149200
7	7	-8.1878405	-8.4006306	-583.2465439
7	6	8.1878405	8.4006306	-928.8669662
8	3	11.5995109	8.1878411	1036.2148895
8	6	-11.5995109	-8.1878411	928.8669586

LOADING 2 WIND FROM RIGHT

JOINT	X FORCE	Y FORCE	BENDING MOMENT
		SUPPORT REACTIONS	
2	13.3343194	11.1952969	-1776.2674866
5	14.7321048	10.3773841	-1890.3127136
8	11.9339367	-21.5726805	-1669.2037506
		APPLIED JOINT LOADS	
1	0.0000219	-0.0000001	-0.0000153
3	0.0000741	0.	-0.0000305
4	-0.0001831	0.0000002	-0.0000458
6	-20.0001414	-0.0000008	-0.0000076
7	-20.0001321	0.0000004	-0.0000229

LOADING 2 WIND FROM RIGHT
JOINT DISPLACEMENTS

JOINT	X DISPLACEMENT	Y DISPLACEMENT	ROTATION
		FREE JOINT DISPLACEMENTS	
1	-3.4057056	-0.0044781	0.0070460
3	-5.3652391	-0.0066073	0.0051562
4	-3.4163731	-0.0041509	0.0048984
6	-5.3745187	0.0110854	0.0037249
7	-3.4295466	0.0086291	0.0094853

PART 1 OF PROBLEM COMPLETED.

```
SELECTIVE OUTPUT
LOADING 1
PRINT DISPLACEMENTS 6,7, DISTORTIONS 8


STRUCTURE SAMPLE STRUCTURE
LOADING 1 UNIFORM ALL BEAMS


JOINT DISPLACEMENTS


JOINT          X DISPLACEMENT   Y DISPLACEMENT       ROTATION

                              FREE JOINT DISPLACEMENTS
  6              -0.0520179      -0.0125412         0.0025205
  7               0.0280228      -0.0089894         0.0009630

MEMBER DISTORTIONS

MEMBER         AXIAL DISTORTION SHEAR DISTORTION BENDING ROTATION
  8              -0.0020754       0.5571141         0.0048139

LOADING 2
PRINT DISTORTIONS 8


STRUCTURE SAMPLE STRUCTURE
LOADING 2 WIND FROM RIGHT


MEMBER DISTORTIONS

MEMBER         AXIAL DISTORTION SHEAR DISTORTION BENDING ROTATION
  8              -0.0092796      -1.2198004        -0.0014313
```

Figure 1.10. Selective output of additional results.

```
MODIFICATION OF FIRST PART - INVESTIGATE ALTERNATE LAYOUT
CHANGES
JOINT COORDINATES
3 X 0. Y 360.
6 X 240. Y 360.
2 X -300.0
ADDITIONS
JOINT RELEASES
2 MOMENT Z
MEMBER RELEASES
2 END MOMENT Z START MOMENT Z
CHANGES
MEMBER PROPERTIES
5 PRISMATIC IZ 600.
LOADING 1
DELETIONS
MEMBER LOADS
8  1
ADDITIONS
MEMBER LOADS
8 FORCE Y LINEAR 0. -0.2
PRINT DATA
```

Figure 1.11. Modification specification and data from internal storage.

```
PROBLEM DATA FROM INTERNAL STORAGE

STRUCTURE SAMPLE STRUCTURE

*  1*TH MODIFICATION OF INITIAL PROBLEM.

STRUCTURAL DATA

TYPE  PLANE FRAME

METHOD STIFFNESS
NUMBER OF JOINTS     8
          MEMBERS    8
          SUPPORTS   3
          LOADINGS   2

JOINT COORDINATES

JOINT       X          Y           Z          STATUS
  1   -240.000    240.000
  2   -300.000      0.                        SUPPORT
  3      0.       360.000
  4      0.       240.000
  5      0.         0.                        SUPPORT
  6    240.000    360.000
  7    240.000    240.000
  8    240.000      0.                        SUPPORT

JOINT RELEASES

 JOINT    FORCE    MOMENT      THETA
NUMBER    X  Y  Z  X  Y  Z      1       2       3
  2                      *      0.      0.      0.

MEMBER  START  END  TYPE          SEGMENT     AX       AY       AZ        IX        IY        IZ        L
  1       2     1   PRISMATIC
                                     1      20.000    0.       0.        0.        0.      200.00
  2       5     4   PRISMATIC
                                     1      20.000    0.       0.        0.        0.      200.00
  3       8     7   PRISMATIC
                                     1      20.000    0.       0.        0.        0.      200.00
  4       1     4   PRISMATIC
                                     1      10.000    0.       0.        0.        0.      300.00
  5       4     7   PRISMATIC
                                     1      10.000    0.       0.        0.        0.      600.00
  6       4     3   PRISMATIC
                                     1      20.000    0.       0.        0.        0.      180.00
  7       7     6   PRISMATIC
                                     1      20.000    0.       0.        0.        0.      180.00
  8       3     6   PRISMATIC
                                     1      10.000    0.       0.        0.        0.      300.00

MEMBER RELEASES

MEMBER         START               END
          FORCE    MOMENT     FORCE    MOMENT
         X  Y  Z  X  Y  Z   X  Y  Z  X  Y  Z
  2                     *                   *

YOUNG-S MODULI
     30000.00 VALUE FOR ALL MEMBERS

LOADING DATA

GIVEN IN TABULAR FORM,WITHOUT LABELS

LOADING 1 UNIFORM ALL BEAMS

TABULATE
        FORCES
        REACTIONS
MEMBER    4 LOAD FORCE  Y UNIFORM         -0.1000     0.          0.         0.
MEMBER    5 LOAD FORCE  Y UNIFORM         -0.1000     0.          0.         0.
MEMBER    8 LOAD FORCE  Y LINEAR           0.        -0.2000      0.         0.

LOADING 2 WIND FROM RIGHT

TABULATE
        FORCES
        REACTIONS
        DISPLACEMENTS
JOINT    6 LOADS                         -20.0000     0.          0.         0.         0.         0.
JOINT    7 LOADS                         -20.0000     0.          0.         0.         0.         0.

     SOLVE
```

Figure 1.11 continued.

STRUCTURE SAMPLE STRUCTURE

MODIFICATION OF FIRST PART - INVESTIGATE ALTERNATE LAYOUT

LOADING 1 UNIFORM ALL BEAMS
MEMBER FORCES

MEMBER	JOINT	AXIAL FORCE	SHEAR FORCE	BENDING MOMENT
1	2	9.8051310	-0.1657431	0.0002613
1	1	-9.8051310	0.1657431	-41.0028343
2	5	35.9713774	0.0000000	-0.0000000
2	4	-35.9713774	-0.0000000	0.0000026
3	8	26.5564461	2.5389214	308.1137924
3	7	-26.5564461	-2.5389214	301.2273331
4	1	2.5388785	9.4721752	41.0028305
4	4	-2.5388785	14.5278245	-647.6807632
5	4	-0.8783489	13.4536604	735.5987930
5	7	0.8783489	10.5463393	-386.7202797
6	4	7.9898897	-3.4172440	-87.9180145
6	3	-7.9898897	3.4172440	-322.1512642
7	7	16.0101073	3.4172501	85.4929314
7	6	-16.0101073	-3.4172501	324.5770836
8	3	3.4172367	7.9898925	322.1512604
8	6	-3.4172367	16.0101070	-324.5770874

MODIFICATION OF FIRST PART - INVESTIGATE ALTERNATE LAYOUT

LOADING 1 UNIFORM ALL BEAMS

JOINT	X FORCE	Y FORCE	BENDING MOMENT
		SUPPORT REACTIONS	
2	2.5388879	9.4721755	0.0002613
5	-0.0000000	35.9713774	-0.0000000
8	-2.5389214	26.5564461	308.1137924
		APPLIED JOINT LOADS	
1	-0.0000094	-0.0000002	-0.0000038
3	-0.0000073	0.0000029	-0.0000038
4	0.0000166	-0.0000027	0.0000229
6	0.0000134	-0.0000002	-0.0000038
7	0.0000202	0.0000005	-0.0000153

MODIFICATION OF FIRST PART - INVESTIGATE ALTERNATE LAYOUT

LOADING 2 WIND FROM RIGHT
MEMBER FORCES

MEMBER	JOINT	AXIAL FORCE	SHEAR FORCE	BENDING MOMENT
1	2	15.1927054	-7.4319584	-0.0026093
1	1	-15.1927054	7.4319584	-1838.5623627
2	5	5.1893103	-0.0000001	0.
2	4	-5.1893103	0.0000001	-0.0000283
3	8	-18.1258667	-29.1056440	-3768.9490356
3	7	18.1258667	29.1056440	-3216.4055176
4	1	10.8948350	12.9365537	1838.5623932
4	4	-10.8948350	-12.9365537	1266.2104950
5	4	-9.9171698	12.8879122	410.2637939
5	7	9.9171698	-12.8879122	2682.8351135
6	4	5.2379517	-20.8117938	-1676.4742584
6	3	-5.2379517	20.8117938	-820.9409943
7	7	-5.2379531	0.8116925	533.5706177
7	6	5.2379531	-0.8116925	-436.1675262
8	3	20.8118558	5.2379527	820.9411697
8	6	-20.8118558	-5.2379527	436.1674728

MODIFICATION OF FIRST PART - INVESTIGATE ALTERNATE LAYOUT

LOADING 2 WIND FROM RIGHT

JOINT	X FORCE	Y FORCE	BENDING MOMENT
		SUPPORT REACTIONS	
2	10.8948308	12.9365745	-0.0026093
5	0.0000001	5.1893103	0.
8	29.1056440	-18.1258667	-3768.9490356
		APPLIED JOINT LOADS	
1	0.0000042	-0.0000207	0.0000305
3	0.0000620	0.0000010	0.0001755
4	-0.0002110	-0.0000002	-0.
6	-20.0001633	0.0000004	-0.0000534
7	-20.0001667	0.0000014	0.0002136

MODIFICATION OF FIRST PART - INVESTIGATE ALTERNATE LAYOUT

LOADING 2 WIND FROM RIGHT
JOINT DISPLACEMENTS

JOINT	X DISPLACEMENT	Y DISPLACEMENT	ROTATION
		SUPPORT DISPLACEMENTS	
2	-0.	-0.	0.0414346
5	-0.	-0.	-0.
8	-0.	-0.	-0.
		FREE JOINT DISPLACEMENTS	
1	-6.9136064	1.7219447	0.0035318
3	-7.5557066	-0.0031233	0.0054063
4	-6.9223223	-0.0020757	-0.0040996
6	-7.5723561	0.0082979	0.0002760
7	-6.9143885	0.0072503	0.0110509

PROBLEM COMPLETED.
TIME USED 26 SECONDS.

Figure 1.12. Modification results.

c. Column 2 hinged at both ends.
d. The moment of inertia of beam 5 doubled.
e. The uniform load on beam 4 changed to a triangular loading with zero intensity at joint 1 and 0.2 kip/inch intensity at joint 4.

To accomplish both analyses in one run, the FINISH statement has to be removed, and the following statements added. Note that the SOLVE THIS PART statement enables the modifications to be read in immediately after the last statement of the original problem. The statements specifying the modifications should be self-explanatory, but they are marked with the letters a through e to identify the changes just described.

```
  MODIFICATION OF FIRST PART - INVESTIGATE
    ALTERNATE LAYOUT
a. CHANGES
  JOINT COORDINATES
  3 X     0. Y 360.
  6 X   240. Y 360.
b. 2 X -300.0          (CHANGE still governs)
  ADDITIONS
  JOINT RELEASES
  2 MOMENT Z
c. MEMBER RELEASES
  2 END MOMENT Z, START MOMENT Z
d. CHANGES
  MEMBER PROPERTIES
  5 PRISMATIC IZ 600.
e. LOADING 1
  DELETIONS
  MEMBER LOADS
  8 1 (first load on member 8 in loading condition 1)
  ADDITIONS
  MEMBER LOADS
  8, FORCE Y, LINEAR 0. -0.2
```

This completes the description of changes. To provide a check on the structure analyzed, we can request a description of the modified structure and loading by using the statement

```
PRINT DATA
```

The program is terminated by the two statements

```
SOLVE
FINISH
```

The affected portion of the STRESS program is shown in Figure 1.11, and the additional results in Figure 1.12. Note that in this case output was obtained only for the quantities requested by the original TABULATE statements, and not for those requested by the SELECTIVE OUTPUT statement.

Chapter 2

DESCRIPTION OF STATEMENTS

2.1 Introduction

The STRESS language has been designed to provide maximum flexibility and ease of use for the engineer. An effort has been made to maintain communication in the engineer's language and yet provide a concise form of input. In a few instances, new terminology had to be introduced. For example, a hinge in a plane frame (zero moment in the plane of the structure) must be described as a member release, moment z.

2.1.1 Order of Statements. The order of statements describing a problem to be solved is arbitrary except for the following:

1. The first statement for a problem must start with the word STRUCTURE.
2. The last statement must be FINISH (or FINISHED).
3. The number of joints must be given before any joint data are given.
4. The number of members must be given before any member data or constants are given.
5. The number of loadings must be given before any loading data are given.
6. The type of structure must be given before member properties are given if the properties include input of the member stiffness or flexibility matrices.
7. Any tabulate statements (2.3.5) appearing after the number of loadings but before any loading headers apply to all loading conditions. Any tabulate statement appearing after a loading statement (anywhere before the next loading header) apply only to that loading condition.

2.1.2 Required Statements. For the specification of a problem the following statements must always be present:

1 STRUCTURE statement
4 NUMBER statements
1 TYPE statement

1 METHOD statement
1 JOINT COORDINATES header statement, plus one statement per joint
1 MEMBER INCIDENCES header statement, plus one statement per member
1 MEMBER PROPERTIES header statement, plus one statement per member
Some load data
1 SOLVE statement
1 FINISH statement
One or more TABULATE or SELECTIVE OUTPUT and PRINT statements

Other statement types will be needed for more complex problems.

2.1.3 Description of Statements. The available statements will now be described in detail. For each statement, the general form is first presented. In the statements forms, capital letters are used for the portions, which must appear in the program exactly as shown; upper and lower case letters are used for describing variable information, and appropriate symbols are used for identifying variable data. Wherever applicable, one or more examples are given for each statement.

2.2 Header Statements

2.2.1 Structure Statement

STRUCTURE Title

This statement initiates a new problem. The title appearing on the statement is used solely for identifying each output tabulation. The title may be any word or group of words suitable for identifying the problem. Example:

STRUCTURE TRIAL DESIGN OF TRANSMISSION TOWER

2.2.2 Loading Statement

LOADING Title

or

LOADING N

The first form of the statement is used to separate groups of individual loads into loading conditions as well as to label the corresponding output tabulations. The second form is used when the statement appears in modification or for selective output to designate the loading condition. During the initial statement of the problem and the addition of new loading conditions during modifications, loading conditions are assigned sequential numbers in the order the loading statements are encountered. The value of N in the second form must then correspond to the assigned number.

If the first word in the title is COMBINATION, the loading condition is taken to be a dependent loading condition, that is, a linear combination of any of the independent loading conditions (see Section 2.5.6). If the first word in the title is not COMBINATION, the loading condition is taken to be independent. Examples:

LOADING LIVE LOAD ON BEAMS
LOADING 3
LOADING COMBINATION DL and LL

2.2.3 Modification Statement

MODIFICATION OF FIRST PART, Title
MODIFICATION OF LAST PART, Title

This statement is used as an output title. It also serves to signal the beginning of a sequence of statements specifying the items to be modified. The first form is used when the modification is to be applied to the problem initially specified. The second form is used when the modifications pertain to the problem just completed. Example:

MODIFICATION OF FIRST PART, INVESTIGATE ALTERNATE LAYOUT
MODIFICATION OF LAST PART, REDUCE STRESS IN COLUMN 15

2.3 Descriptors

2.3.1 Size Descriptors. The following four statements must appear in every problem, followed by an integer number (no decimal point):

NUMBER OF JOINTS N
NUMBER OF SUPPORTS N
NUMBER OF MEMBERS N
NUMBER OF LOADINGS N

These statements are self-explanatory. The NUMBER OF LOADINGS is the number of loading conditions (not individual loads) for which a solution is desired.

2.3.2 Type Statement. The structural type is given in a statement with the first word TYPE followed by a description, which must be one of the following:

TYPE PLANE TRUSS
TYPE PLANE FRAME
TYPE PLANE GRID
TYPE SPACE TRUSS
TYPE SPACE FRAME

Only one type statement can be given in a problem specification.

2.3.3 Method Statement.

METHOD STIFFNESS

This statement must appear in this form. Additional methods of analysis will be incorporated into STRESS as they are developed and may include the flexibility method, statically determinate analysis for forces, limit analysis, and so on.

2.3.4 Tabulate Statement. Tabulated output may be requested for all joints and members by one or more statements starting with the word

TABULATE

followed by any of the following descriptors, singly or in a string:

FORCES	member forces at the ends of each member.
REACTIONS	support reactions and joint loads as back-substituted from the member forces. The latter constitutes a useful statics check to evaluate the roundoff error incurred in the computation.
DISTORTIONS	member distortions, the sum of the strains between the member ends. All strains are given, that is, axial and shear distortions and rotations between end tangents.
DISPLACEMENTS	joint displacements.
ALL	all of the above four.

The words MEMBER or JOINT may be placed before the appropriate descriptors to increase clarity.

If the member distortions of a released member are printed, they include the distortion of the release (for example, rotation of a hinge) as well as those of the member. Examples:

TABULATE FORCES
TABULATE JOINT REACTIONS, MEMBER FORCES
TABULATE ALL

2.3.5 Termination Statements. The following statements are available:

SOLVE

For the particular problem specification, a solution will be performed with the requested tabulation of results using the data given up to this statement. The problem will then be terminated. Any additional input will be scanned for correctness of form.

SOLVE THIS PART

Same as SOLVE, except that upon completion selective output and modification may be undertaken. If an input or execution error occurs, the modification data following this statement are only scanned.

FINISH or FINISHED

ndicates the end of problem. The time used for the solution is printed ut. The next problem is then begun.

STOP

erminates processing, and is placed after the last problem in a series rocessed together.

2.3.6 Selective Output Statements. A selective output mode is also vailable for obtaining only particular data instead of the total possible utput of a type requested by TABULATE. This mode is entered by the ingle statement

SELECTIVE OUTPUT

laced after the SOLVE THIS PART statement. It is followed by the iden-ification of the loading condition, for example,

LOADING 2

The printing for selective output is requested by a statement starting ith PRINT, followed by one of the four output descriptors listed under 'ABULATE and a string of member or joint numbers. Additional de-criptors and numbers may be strung onto the statement or included in a eparate statement. Example:

PRINT FORCES 10, 12, 14, REACTIONS 1, 2, 3

n addition, the statement

PRINT DATA

ill give a listing of all of the current data upon which the system is op-rating. This is useful in stating concisely a problem that may have be-ome obscured by many modifications. The SELECTIVE OUTPUT state-nent is not needed before a PRINT DATA statement.

2.3.7 Modification Descriptors. There are three single-word state-nents in this group:

ADDITIONS
CHANGES
DELETIONS

These statements are used to indicate the type of modification to be per-ormed on the data and descriptors by the statements immediately fol-owing. A modification descriptor statement is effective until another tatement in this group or a termination statement is encountered. De-etion of a joint implies deletion of the members framing into that joint s well as the corresponding joint and member loads. Implied deletions nay also be explicitly stated. Items placed after CHANGES are written n the same form as the original data, except as described below for loads. n all cases the changed data replace the original values.

A second level of modification is required for changes in a loading condition by any type of modification of individual loads. The first statement must be CHANGES, followed by the statement LOADING N (see Section 2.2.2), in turn followed by the appropriate CHANGES, ADDITIONS or DELETIONS statement. As an example, the following set of statements will delete the second load statement previously given for joint 3 in the first loading condition:

CHANGES
LOADING 1
DELETIONS
JOINT 3 LOAD 2

If the load statement number is omitted it is taken as 1. Loading statement numbers for a joint or member in a loading condition are assigned according to the input order neglecting previously deleted statements. Specifying a LOADING under DELETIONS indicates deletion of all loads in that loading condition.

2.4 Structural Data

As described in Sections 1.7 and 1.8, the structural data for joints and members are given by a header statement for each type of data, followed by a tabular statement for each member or joint, giving the member or joint number, the labels identifying the numerical data, and the data them selves. Certain alternatives for these forms are discussed in Section 2.

Labels are indicated by one, two, or three words, such as X, FORCE X, or END FORCE X, depending on context. When a series of labels is given, leading words need not be repeated (that is, FORCE X, Y is equivalent to FORCE X, FORCE Y). In some cases (for example, releases), the labels themselves constitute the data.

In the following sections describing the structural and load data statements, both the tabular header statement and the corresponding tabular statement will be shown and described. The symbol J will be used for joint numbers and M for member numbers. Since these numbers are integers, they require no decimal point. All data values must be given with a decimal point except those for M and J. Zero data values need not be entered if the corresponding labels are omitted. The number of labels must be equal to the number of pieces of numeric data, except for variab section member properties data. The symbols α, β, and θ, where they appear in the following, indicate appropriate numeric data.

2.4.1 Coordinate Statement

JOINT COORDINATES
J X α_1 Y α_2 Z α_3 Status

Joint coordinates are given in an arbitrary orthogonal global coordinate system. The status indicates whether the joint is a support joint or unrestrained. If the status is omitted, the joint is assumed to be free.

SUPPORT or S indicates the joint is a support, and FREE or F indicates the joint is free. The latter designation is needed only in changing the status of a joint. The Z coordinate may be omitted for plane structures. Examples:

10 X 15.0 Y 25.0 Z 35.0 SUPPORT
12 X 20.0 S (Y is zero)

2.4.2 Joint Release Statement

JOINT RELEASES
J FORCE X Y Z MOMENT X Y Z, θ_1 θ_2 θ_3

The joint release statement is used to indicate the condition of a support joint that is not fully fixed against displacement in all possible component directions. Deviations from full fixity are given by listing the components that are free to move, that is, are released. Labels are the data for the presence of releases in an orthogonal coordinate system which need not be parallel to the global system. As many labels as necessary may be given. The order of the labels is arbitrary. The orientation of the release coordinate system from the global system is given by the angles θ_1, θ_2, θ_3, as shown in Figure 1.4. These angles are given in decimal degrees and have no labels. For a plane structure only one angle is needed. Examples:

3 FORCE X MOMENT Z (this is the customary horizontal roller in a plane frame)
4 FORCE X Y 45.0 45.0 0.0

2.4.3 Member Incidence Statement

MEMBER INCIDENCES
M JA JB

Joint JA is the START of member M, and joint JB the END. This statement is also used to define the member x axis. The direction of the x-axis is from the starting joint to the end joint. JA and JB are integer numbers.

2.4.4 Member Release Statement

MEMBER RELEASE
M END FORCE X Y Z MOMENT X Y Z START FORCE X Y Z
MOMENT X Y Z

A member release is used to designate that a force component at one end of a member is always zero. The member force and moment releases are specified in the local member coordinate system. The positions of releases can only be at the member ends, and therefore releases at any other point require the insertion of a joint at that point. As with joint releases, the presence of the labels in the data, and the words, END, START, FORCE and MOMENT may, but need not, be repeated. Any number of releases may be specified in one or more statements as long as the member is not made unstable. Examples:

> 10 END MOMENT Z START MOMENT Z (the bar is hinged at both ends)
> 10 START FORCE Y MOMENT Z (the bar has a hinge and roller that is, at one end it transmits axial forces only)

2.4.6 Member Properties Statement

> MEMBER PROPERTIES, List of common properties
> M Properties not included in the list of common properties

The possible forms for specifying properties are described in this sec tion. In each case a word describing the type of member properties mus be given with appropriate labels first. This word is then followed by the data. Properties common to all members in a table may, but need not, appear in the table heading. Note that the word PROPERTIES is used onl in the tabular heading. The table heading may have the form

> MEMBER PROPERTIES

or may include the type of properties, such as

> MEMBER PROPERTIES, PRISMATIC

or may include in addition data common to all or most of the members to be listed in the table. If the member type is given in the heading, it may not be repeated in the subsequent entries in the table. If data values are given for some or all of the properties for a single member with differen values than those given in the header statement, the latter values will be used.

Area and moment of inertia labels may be given either as one- or two-word labels. One-word labels may be any of the following:

> AX for the normal cross-sectional area
> AY for the effective shearing area in the y direction
> AZ for the effective shearing area in the z direction
> IX for the torsional constant
> IY for the moment of inertia about the y axis
> IZ for the moment of inertia about the z axis

For two-word labels, the first word may be either A or AREA and I or INERTIA and the second word X, Y, or Z. The shearing flexibility is taken to be zero if AY or AZ are either not given or given as zero. The following type and forms of properties may be used:

1. PRISMATIC AX α_1 AY α_2 AZ α_3 IX α_4 IY α_5 IZ α_6 BETA β_7

This form may be used to specify the properties of straight prismatic members of constant section between the member ends. BETA signifies the angle β for member rotation defined in Section 1.3. Only the properties that enter into the analysis for the appropriate structural type (see Table 1.1, p. 7) need to be given.

2. VARIABLE, N SEGMENTS, BETA β AX α_1 AY α_2 AZ α_3 IX α_4 IY α_5 IZ α_6 L α_7 α_8 α_9 . . .

If BETA is not given, β is taken as zero.

A straight member with stepwise variation of section properties is designated by VARIABLE. The number of segments for which the properties are specified is denoted by N, an integer; L designates the length of the segment to which the properties pertain. A maximum of 9 segments per member may be used. Only one value of β, however, is permitted. Labels are given once to show the existence and order of the data. The data for each of the segments are given in order from the START to the END of the member. The number of pieces of data (not counting BETA and the value β) expected is then N times the number of labels.

Since the label BETA corresponds to only one piece of data, while the other labels correspond to N pieces of data, an ordering convention must be adopted which separates β from the other properties. The convention is that if β is given, BETA and the value β are to be given first.

If any data values are given in a tabular header, the number of segments must also be given in the tabular header and not given in the table entry statements. If no data values are given, the number of segments should not be given in the heading. If some data is specified in the heading and some in the table entries, N pieces of data are expected for each label, (except BETA) with the ordering rules above applying.

3. FLEXIBILITY GIVEN, BETA β, Data
4. STIFFNESS GIVEN, BETA β, Data

The flexibility or stiffness matrix of the member, consisting of JF $\times$ JF (see Table 1.1, p. 7) entries, is given by rows. The order of matrix components in each row is the order given in Table 1.1 for each type. If the type is changed by modification, these matrices must then be changed. With subscripts representing directions, the first row of a stiffness matrix for a member in a plane frame would be

$$k_{11}\ k_{12}\ k_{16}$$

As before, if a β angle is present, it must be labeled and placed first. If the member contains releases, the user has the option of either specifying the stiffness or flexibility matrix for the "fixed-fixed" member and the corresponding releases, or giving the modified matrices and not using the RELEASE statement.

5. STEEL, Section name, BETA

This form has not yet been included in the STRESS system. When it is included, data for most standard steel sections will be available. Examples:

```
10 STEEL 21WF112
11 STEEL 18WF55, BETA 30.
```

2.4.7 Constants Statement

CONSTANTS Name, α_1, N_1, . . . N_n, α_2, N_1, . . . N_m
CONSTANTS Name, α_1, ALL
CONSTANTS Name, α_1, ALL BUT, α_2, N_1, . . . N_n

Constants relating to the members can be given individually or in string form as shown. Any of the following names may be given:

E - Young's Modulus
G - Shear Modulus (if not given, G = 0, 4E)
CTE - Coefficient of thermal expansion
DENSITY - Material density

(CTE and DENSITY have no use until additional capabilities are included in the processor.) The α's are decimal numbers with not more than 8 digits before the decimal point. All N's following a value are the member numbers to which the given value applies. If the same value of a constan applies to all members, ALL may be written after the value. If only a few members have constants with different values, ALL BUT will assign the more common value to all members, and then overwrite these with the different values when given. Examples:

CONSTANTS E 30000. ALL BUT 10000. 10, 12, 13, G 5000.
10, 12, 13

2.5 Load Data

As described in Section 2.2.2, all load data following a LOADING heade are considered to pertain to a single loading condition. The number of loads in a loading condition is arbitrary. Labels are used to identify data in the same manner as for the structural data.

2.5.1 Joint Loads Statement

JOINT LOADS
J FORCE X α_1 Y α_2 Z α_3 MOMENT X α_4 Y α_5 Z α_6

JOINT LOADS are concentrated forces and moments acting on the joints of a structure. Six labels are shown for the six possible components of joint loads. Only the components actually present need be given. Any or all of these components may be given in a single statement, and more thar one statement for a particular joint may be given in a loading condition. JOINT LOADS may not be specified for support joints unless JOINT RELEASE has been specified for those joints. The coordinate system for joint loads and displacements is the global system with loads and displace ments positive when in the positive direction. Examples:

8 FORCE X 10. Y -10. MOMENT Z 200.
9 MOMENT Z -100. FORCE Y 30.

2.5.2 Joint Displacements Statement

JOINT DISPLACEMENTS
J DISPLACEMENT X α_1 Y α_2 Z α_3 ROTATION X α_4 Y α_5 Z α_6

Joint displacements are prescribed joint motions and may therefore be used only at support joints. This statement is similar to joint loads. For ease of input, the label DISPLACEMENT can be omitted if the labels X, Y, and/or Z are given before the word ROTATION. Rotation quantities are given in radians. Example:

2 DISPLACEMENT X Y .2 .4 ROTATION Z .003

2.5.3 Member Loads Statement

MEMBER LOADS
M Direction, Type, Labels, Data

A member load statement specifies one force or moment component acting between the member ends.

Direction is a label that must be one of the following:

FORCE X
FORCE Y
FORCE Z
MOMENT X
MOMENT Y
MOMENT Z

Applied moments are specified in terms of the axes about which they act.

Three types of member loads are acceptable and are shown with the appropriate labels:

CONCENTRATED P α_1 L α_2
UNIFORM W α_1 LA α_2 LB α_3
LINEAR WA α_1 WB α_2 LA α_3 LB α_4

where

P = load intensity
L = distance from member START to point of application (see MEMBER INCIDENCES)
W = load intensity, uniform between LA and LB
LA = distance from START to beginning of load
LB = distance from START to end of load
WA = load intensity at beginning of load
WB = load intensity at end of load

If both LA and LB are not given or equal to zero, it is assumed that the load extends over the entire length of the member. Examples:

14 FORCE Y CONCENTRATED P -20. L 50.
15 MOMENT Z UNIFORM W 1.

2.5.4 Member Distortion Statement

MEMBER DISTORTIONS
M DISTORTION X α_1 Y α_2 Z α_3 ROTATION X α_4 Y α_5 Z α_6

This statement is similar in use to JOINT DISPLACEMENTS. Distortions are displacements between the member ends, positive if the displacement from the member START to the member END is in the positive local coordinate direction. Rotation distortion is the angle in radians between the tangents at the ends if the member were free, with the sign defined similarly. Example:

13 DISTORTION X .1 Y .2 Z .4

2.5.5 Member End Load Statement

MEMBER END LOADS
M END FORCE X α_1 Y α_2 Z α_3 MOMENT X α_4 Y α_5 Z α_6
START FORCE X α_7 Y α_8 Z α_9 MOMENT X α_{10} Y α_{11} Z α_{12}

In cases where the member flexibility or stiffness matrix is given or the loading pattern is different from the three types discussed under MEMBER LOADS, the processor does not have sufficient information to compute the fixed-end forces and moments from member loads. It is then necessary to input the member end loads ("fixed-end forces") computed outside the processor. These forces are given acting on the member ends. If releases have been specified for the member, the end forces must correspond to the properties and action of the released member. Example:

4 END FORCE X Y MOMENT Z −10. −20. −300. START
FORCE X Y MOMENT Z 10. 20. −300.

2.5.6 Combine Loadings Statement

COMBINE N_1 α_1, N_2 α_2

The loading header under which this statement occurs must contain the word COMBINATION as the first word in the title. Any number of combine statements may be used. The N's are the sequential loading condition numbers (see Section 2.2.2). The α values are decimal numbers specifying the coefficients to be used in summing the contributions of the appropriate loading conditions. A combination loading is a linear combination of independent loadings, all of which must be specified prior to the combination. In other words, if N_i is the sequential number of the combination loading, all N's must be less than N_i. Example:

COMBINE 3 1. 4 .25 5 3.7

2.6 Modifications

If the termination statement SOLVE THIS PART is used after a problem specification, the problem may be easily modified for subsequent resolution. After the execution of the initial problem (including SELECTIVE

OUTPUT), a modification header (Section 2.2.3) restores the appropriate data for modification. Modification of the data continues until a SOLVE or SOLVE THIS PART statement is encountered, and may in turn be followed by additional modifications.

Modifications are specified by the statements previously described, with the following exceptions. First, additional information is needed to specify what kind of modification is intended by a statement. The modification descriptors (Section 2.3.7) are used only for this purpose. A modification descriptor specifies the kind of modification performed by the statements which follow it until either the next modification descriptor or a termination statement is encountered.

For DELETIONS, no decimal data need be given where indicated in the statement explanation. Only a member or joint number need be given in a statement following the table headers MEMBER INCIDENCES or JOINT COORDINATES. Deletion of a joint implies deletion of all of its joint loads, all members framing into the joint and all of their corresponding member loads. Deletion of a member implies deletion of all of its loads. These implicit deletions are, however, performed only after the termination statement. Therefore, if an additional member or joint is specified during a modification with the same number as a deleted member or joint, the implied deletions must be done explicitly in order to be properly executed.

The TYPE and NUMBER statements may appear only under CHANGES, with their form as shown in Section 2.3.

Almost all structural data may appear under any of the modification descriptors. The ADDITIONS for member and joint data pertains only to the added members and joints. Additional data for any existing member or joint are considered to be a change of existing data. Constants may not be deleted.

A loading header appearing under CHANGES or DELETIONS must have a loading number corresponding to its sequential input order. The title may be omitted. Deletion of a loading condition indicates deletion of all loads. Changes in a loading condition indicate modification of the TABULATE statements and/or its loads. Therefore, a second level of modification is required. The statement LOADING N following CHANGES sets the loading sequence number on which to operate. Joint and member load statements (loads or distortions, and so forth) under CHANGES or DELETIONS must also contain a load statement sequence number on the member or joint in the loading condition involved. For example, following the statement DELETIONS the statements

```
JOINT LOAD
2 3
```

deletes the data for the third joint load statement in sequence specified for joint 2. If the load statement number is omitted during deletions, it is taken as one, that is, the first load is deleted.

A loading header under ADDITIONS initiates a new loading condition and need not contain a value for N. All the following load statements, as well as load statements appearing under ADDITIONS during CHANGES of an existing loading condition, need not have a statement number after the load type (load, displacements, distortions, end load).

Modification of both releases and tabulate requests involves the newly stated components and those previously stated, if any. Under ADDITIONS, the newly stated components are added to the previously stated components, if any. Under DELETIONS, the newly stated components are deleted from the existing set of components for the joint, member, or loading condition. In the CHANGES mode, all previously stated components are deleted and the newly stated components added.

Changing any quantity from a nonzero value to zero requires specifying zero. Only changed data need be specified in a statement if the form with labels is used. This facility allows ease and brevity in modifications.

2.7 Alternate Input Forms

In order to facilitate input of member and joint data, three alternate input forms are provided. The first two alternates may be found convenient for entering data for a particular member or joint, while the third may be used to make the input more legible or to reduce the number of statements needed for specifying modifications.

2.7.1 Rearrangement of labels and data. In all statements requiring both data labels and data values, the labels serve only to indicate the order of the data values. Labels need not be placed next to the data values they refer to, as long as the data values are given in the same sequential order as the labels. Thus the last two of the following three statements are equivalent:

```
MEMBER PROPERTIES, PRISMATIC
3 AX 10.0 AY 20.0 IZ 400.0
3 AX AY  IZ 10.0 20.0 400.0
```

Grouping of labels, with liberal use of DITTO's or DO's, can simplify input of long tables of member or joint data, both for structural properties or loads.

The order of labels and data also is arbitrary and need not correspond to the order given.

2.7.2 Fixed order of data without labels. To further simplify input, at the risk of greatly reduced legibility and increased number of values, data values can be entered without labels, provided that a fixed order of data is maintained. For each statement, the fixed order is that of the labels shown for the appropriate statement. It should be noted that the labels shown are those for the general case, that is, a space frame. This order is maintained so that zeroes must be inserted for unused values. If the example of the previous section is used, the properties of member 3 may be entered as

3 10.0 20.0 0.0 0. 0 0.0 400.0

An example of equivalence for load data is

MEMBER LOADS
17 FORCE Y UNIFORM W -0.1 LA 10.0 LB 20.0
17 FORCE Y UNIFORM W LB LA -0.1 20.0 10.0
17 FORCE Y UNIFORM -0.1 10.0 20.0

For convenience in the input of member properties with fixed order of data, the value β is to be given last and therefore can be omitted if equal to zero. For example, a member stiffness matrix with a β angle must be given in the following form:

STIFFNESS GIVEN, matrix data, β

For variable section properties, 7 times N pieces of data are expected, plus a value of β if not zero.

Table 2.1. Statement Variations

Tabular Header and Tabular Statement	Equivalent Descriptive Statement
JOINT COORDINATES J Labels, Data, Status	JOINT J COORDINATES Labels, Data, Status
JOINT RELEASES J Labels, Angles	JOINT J RELEASES Labels, Angles
MEMBER INCIDENCES M JA JB	MEMBER M GOES FROM JA TO JB or MEMBER M FROM JA TO JB
MEMBER RELEASES M Labels	MEMBER M RELEASES Labels
MEMBER PROPERTIES M Properties, Labels, Data	MEMBER M Properties, Labels, Data
JOINT LOADS J Labels, Data	JOINT J LOADS Labels, Data
JOINT DISPLACEMENTS J Labels, Data	JOINT J DISPLACEMENTS Labels, Data
MEMBER LOADS M Direction, Type, Labels, Data	MEMBER M LOADS Direction, Type, Labels, Data
MEMBER DISTORTIONS M Labels, Data	MEMBER M DISTORTIONS Labels, Data
MEMBER END LOADS M Labels, Data	MEMBER M END LOADS Labels, Data

2.7.3 Descriptive Statement Forms. As an alternate to the tabular input for member or joint data, descriptive forms are provided for all tabular statements. The descriptive forms may be used interchangeably with the corresponding tabular forms. It should be recalled that tabular input is terminated when the first item on a statement is not an integer number, and that more than one tabular header of a given type may be used. The primary use of the descriptive form is to increase the legibility of the input, or to simplify modification specifications, where generally only one or a few statements are modified at a time.

The right-hand column of Table 2.1 gives the descriptive statements which are equivalent to the header and tabular statement combinations shown in the left-hand column.

It should be noted that the words INCIDENCES and PROPERTIES may not be used in the descriptive statement forms. The descriptive form of member or joint deletion is merely (under DELETION mode) MEMBER M or JOINT J.

Chapter 3

USE OF STRESS

3.1 Size Limitations

The STRESS processor incorporates several unique computational methods. One of these is an automatic memory control scheme which allows a freedom in problem size while maintaining efficiency to a large degree. As a result, it is difficult to state fixed limits on problem size. The following are approximate statements and refer to the probable limits at expected bottlenecks for the total process. These limits can be changed by further partitioning the data arrays and/or the STRESS processor program.

For an IBM 709/7090/7094 data processing system with 32,768 storage locations and using a 4-link partitioning of the processor, the following estimates are offered:

1. No more than 1500 statements in one initial problem specification.
2. $NB \times (16 + 3 \times JF) + NJ \times NLDS \times JF \leq 17000$
3. $NJR \leq 60$

where:

NJR = total number of released force components at support joints
JF = number of degrees of freedom per joint
NB = number of members
NJ = number of joints
NLDS = number of loading conditions

All of the listed requirements must be met. If CONSTANTS are not given for the individual members, the 16 in requirement 2 is replaced with a 12. There is no limit on the number of modifications.

3.2 Card Punching

With the following nine rules, a STRESS problem can easily be punched on an IBM 26 card punch. The form of the input is free field, that is, there is no particular column on a card where a piece of information must start or end. There are, however, the following few restrictions of form:

1. Only columns 1 to 72 on a card are used.
2. A statement may start anywhere on the card.
3. Once a statement has been started, no more than 5 blanks may be left between words and/or numbers.
4. Only the first 6 letters in a word are used. Additional letters are ignored by the system and may be omitted.
5. An * in card column 1 denotes a comment card, which is printed on the output but otherwise ignored. Any number of comment cards may be used.
6. A $ in card column 1 denotes that this card is a continuation of the statement of the previous card. No separation of word parts is allowed, and care must be used to follow rule 3 at the end of the previous card. In the case where a long word does not fit on the end of the previous card, rule 4 can be applied. Any number of continuation cards may be used to complete a statement.
7. All numeric data values which can attain noninteger values must contain a decimal point. Integer numbers such as joints, members, loadings, and so forth, must not have a decimal point.
8. DITTO or DO may be used to repeat a word or label from the last statement of the same type.
9. Commas may be used to separate words or may be omitted. At least a comma or a blank must be used for separating words and/or numbers. A comma counts as a word when rule 3 is applied.

3.3 Running at the M.I.T. Computation Center

The computer used at M.I.T. is an IBM 7094 with many special features normally not available on other machines. Only one of these, however, is utilized by the STRESS processor. This is the interval timer employed to time individual problems.

After a deck of cards has been punched which represents a problem (from STRUCTURE to FINISH), the deck must be combined with a processor. This is done by obtaining a STRESS program starter deck and using the card arrangement shown in Figure 3.1. Any number of problems may be included in a deck. The ID card contains an * in column 1 and a problem number under which the program is to be run. A request card must be time-punched and filled out, including as a special instruction "STRESS."

3.4 Output

STRESS output request statements have been described in Section 2.3. Output is given in tabular form with column headings that completely identify the data.

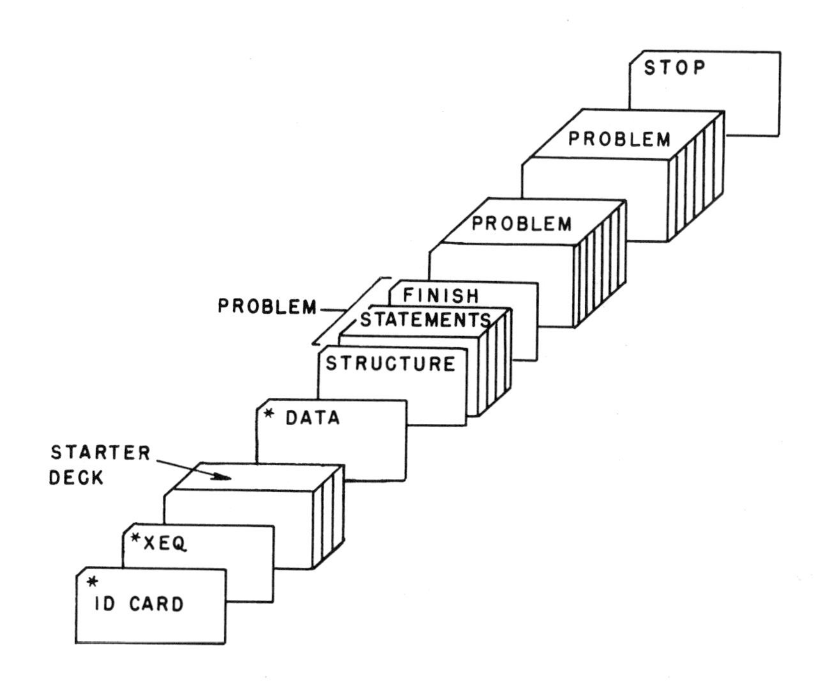

Figure 3.1. Input deck arrangement.

In order to interpret properly the sign of a result, it is necessary to construct mentally coordinate axes at that point. Member results are given in local coordinates, joint results in global coordinates. Both are right-hand orthogonal systems. In all cases, positive quantities denote forces or moments in the positive direction (see Figure 1.1, p. 4). As an example, a compressive axial force is shown as positive at the member start (direction of force same as orientation of the member) and negative at the member end. It must be noted that for members, the x-axis direction is from start to end, and the y axis is either given by a β angle or, for the plane structure, is in the plane by the right-hand rule.

3.5 Error Messages

All reasonable effort has been made to check input to STRESS for errors and provide diagnostic information essential to the user. There are three categories of error messages:

a. Messages pertaining to errors in the statements, printed immediately underneath the erroneous statements.
b. Messages pertaining to errors of omission or commission on the problem as a whole, printed after the SOLVE or SOLVE THIS PART statement.
c. Messages identifying errors detected during the execution of the problem.

In addition, certain messages are printed when major steps in the solution have been completed.

Although all error messages are believed to be self-explanatory, they are presented here for completeness. In the messages in the next sections, underlined items denote examples of variable information supplied by the processor.

3.5.1 Errors in Input Statements. The following messages are printed immediately below the erroneous statement:

Message	Explanation
THE WORD * MEXMER * CANNOT BE TRANSLATED	The statement contains an unacceptable word (probably a misspelling).
UNACCEPTABLE DITTO	One of the DITTO's in the statement is either in the wrong place or should not be present. Compare statement with previous one.
TOO MANY BLANKS	More than 5 blanks between two items.
FIXED POINT NUMBER NOT FOUND	A fixed-point number missing from statement. Check form of statement.
FLOATING POINT NUMBER IN WRONG PLACE	Check form of statement.
FIXED POINT NUMBER IN WRONG PLACE	Check form of statement.
STATEMENT NOT ACCEPTABLE	Check form of statement.
JOINT NUMBER GREATER THAN NUMBER OF JOINTS SPECIFIED MEMBER NUMBER GREATER THAN NUMBER OF MEMBERS SPECIFIED LOADING NUMBER GREATER THAN NUMBER OF LOADING SPECIFIED	For these three messages, check that appropriate size descriptor precedes erroneous statement.
TYPE OF MODIFICATION NOT SPECIFIED	The MODIFICATION statement has not been followed by an ADDITIONS, CHANGES, or DELETIONS statement.
MODIFICATION NOT ACCEPTABLE	Check Section 2.6.
TYPE OF STRUCTURE NOT ACCEPTABLE	Check Section 2.3.2.
METHOD SPECIFIED NOT AVAILABLE	Check Section 2.3.3.

3.5.2 Errors in Consistency of Problem. If the problem specified is not executable, the following messages are printed:

TYPE OF STRUCTURE NOT SPECIFIED
METHOD OF SOLUTION NOT SPECIFIED
NO LOADS SPECIFIED
NUMBER OF JOINTS NOT SPECIFIED
NUMBER OF SUPPORTS NOT SPECIFIED
NUMBER OF MEMBERS NOT SPECIFIED
NUMBER OF LOADINGS NOT SPECIFIED
NUMBER OF JOINTS GIVEN NOT EQUAL TO THE NUMBER SPECIFIED
NUMBER OF SUPPORTS GIVEN NOT EQUAL TO THE NUMBER SPECIFIED
NUMBER OF MEMBERS GIVEN NOT EQUAL TO THE NUMBER SPECIFIED
NUMBER OF LOADINGS GIVEN NOT EQUAL TO THE NUMBER SPECIFIED
NUMBER OF MEMBER PROPERTIES GIVEN NOT EQUAL TO THE NUMBER OF MEMBER GIVEN

UNACCEPTABLE STATEMENTS PRESENT	Problem cannot be executed because of errors in the input statements which have been individually identified.
STRUCTURAL DATA INCORRECT	The actual error has been identified as an unacceptable statement.
LOADING DATA INCORRECT	The actual error has been identified as an unacceptable statement.
THE 10TH LOAD DATA FOR JOINT 11 ARE INCORRECT	A joint displacement has been specified at a free joint or a joint load at a nonreleased support joint.

PROBLEM INCORRECTLY SPECIFIED

3.5.3 Errors in Execution. If an error is encountered during execution, the appropriate messages are printed:

MEMBER 1 SINGULAR FLEXIBILITY MATRIX
OVERFLOW INVERTING FLEXIBILITY MATRIX, MEMBER 2
MEMBER 5 UNSTABLE, TOO MANY RELEASES
SINGULARITY DUE TO JOINT RELEASES

8TH DIAGONAL ELEMENT OF STIFFNESS MATRIX IS SINGULAR	The structure is probably unstable. Check MEMBER INCIDENCES statements, or better, sketch structure from the MEMBER INCIDENCES data.

MEMBER TYPE AND LOAD TYPE ARE INCOMPATIBLE, MEMBER 15 LOAD TYPE 2
LOAD ON MEMBER 16 LOADING CONDITION 2 INCOMPATIBLE WITH STRUCTURE TYPE
COMPUTED LENGTH OF MEMBER 3 NOT EQUAL TO THE SUM OF THE SEGMENTS
THE 12TH SEGMENT HAS BEEN ALTERED TO 13.14

The last two messages are provided when a VARIABLE member has been improperly specified, but the execution is not interrupted.

With the preceding exception, errors in consistency or in execution cause termination of execution, and the message

PROBLEM FAILED is printed.

If SOLVE THIS PART was used, the message

EXECUTION DELETED - FOLLOWING STATEMENTS SCANNED ONLY

is printed, and all statements up to FINISH are scanned for correct form only. Any errors located in these statements are identified by the messages listed in Section 3.5.1.

3.5.4 Monitoring Messages.

PART 6 OF PROBLEM COMPLETED	Printed after each execution initiated by a SOLVE THIS PART statement.
PROBLEM COMPLETED	Printed when an entire problem has been completed.
TIME USED 4 SECONDS	The total time, truncated to the nearest second, used to process the entire problem is printed.

Chapter 4

DESCRIPTION of the STRESS SYSTEM

This chapter gives a brief description of the internal structure of STRESS. It is assumed that the reader has some familiarity with computers and with conventional compilers, such as FORTRAN or MAD.

STRESS is neither a general-purpose program nor a compiler in the conventional sense but combines certain features of both. To the user, it resembles a compiler, in the sense that it consists of an input or source language unintelligible to the computer and a system or processor that processes the input language to produce the desired results. However, there is no "object program" produced, so that a run on STRESS is equivalent to the "compile-and-go" mode for conventional compilers.

The STRESS system performs four basic functions, namely, input, compilation, execution, and modification. These functions are briefly described in the following sections.

4.1 Input

The first phase of STRESS is the input phase. It is initiated by a STRUCTURE statement and suspended by a SOLVE or SOLVE THIS PART statement. As each statement is read, the logical fields on the card (that is, words or numbers, separated by one or more blanks or commas) are successively scanned from left to right, and matched against tables, or, in the case of numbers, converted to the proper internal representation. After the statement has thus been decoded, the data contained in it are stored, parameters specified by the labels are set, and "flags" needed for later checks are set. For example, when a TYPE statement is read, the following two words (the name) are translated into a numerical code from 1 to 5, which becomes the parameter ID, used by the remainder of the program to distinguish between the various types; also, a flag is set signifying that this statement, which must be present in each problem, has indeed been encountered.

If a portion of the statement is unacceptable for any reason (such as misspelling, use of wrong word, or wrong sequence of words), decoding of the statement is terminated, an identifying error message is printed (see

Section 3.5.1), and a flag is set. This flag will prevent execution of the problem without interfering with the processing of the remaining input. If, for any reason, the execution of a portion of a problem must be interrupted, the remaining input for that problem will be scanned for possible errors. Throughout the system, a great effort is made to provide as much diagnostic information as possible. Thus, even on the largest problems, it is normal to "debug" the user's program in one pass, or at most a very few passes.

4.2 Compiling

The second phase of the system performs all the editing, checking, and compilation functions. First, certain blocks of input data are renumbered for more efficient execution. Next, additional checks are performed, and all the flags set by the input and compilation phases are examined simultaneously. This check includes the detection of all errors of commission or omission which may be "fatal," that is, preclude the proper execution of the problem. If any errors are detected, diagnostic information is printed out (see Section 3.5.2), and control returns to the input phase. Any succeeding statements are then scanned until a FINISH statement is encountered.

If the problem is executable, the compilation functions are performed. Since the size of all the data arrays that will be needed are known at this point, STRESS assigns the exact number of storage locations for each array. This is equivalent to a computed DIMENSION statement in FORTRAN. However, these assignments are made only symbolically, as STRESS uses a dynamic memory allocation procedure, and a particular array is in core storage only when actually needed.

The compilation of subroutines for the execution of a problem is handled in different versions of the system either by setting up a linkage of CHAIN segments or by a dynamic allocation scheme similar to that used for data.

The compilation phase automatically calls in the next phase, execution.

4.3 Execution

The third phase is the actual execution. It should be noted that, as a result of the compilation phase, the program in memory looks essentially like a special-purpose structural analysis program written specifically for the particular problem being run. Furthermore, the configuration of memory may change continuously, as data arrays and program segments are brought in or temporarily stored away, depending on the size and complexity of the problem.

In essence, during execution, a control program calls in and executes the appropriate subroutines, checks for fatal errors (such as singularity of a stiffness matrix), and terminates execution if an error occurs (see

Section 3.5.3). In general, termination caused by errors is delayed as much as possible, so as to scan the largest number of potential errors. An unexpected termination is handled in the same way as in the compilation phase, that is, by returning to the input phase. The execution generally proceeds along the following steps:

1. Member computations. The input specifications of member properties are transformed into appropriate stiffness or flexibility matrices. Member releases, if any, are incorporated by modifying these matrices.
2. Structure computations. By using a network formulation,[1] a compact representation of the stiffness or flexibility matrix of the structure is generated. Joint releases are taken into account.
3. Loading computations. The input specifications of applied loads, displacements, or distortions are converted to equivalent joint loads.
4. Solution and back-substitution. The resulting equations are solved and back-substitutions performed.
5. Output. The desired results, specified by the TABULATE statements, are computed and printed.

After execution is completed, the system again returns to the input phase.

4.4 Modifications

Upon return from the execution phase, the input phase resumes the reading and decoding of statements. If the compilation and execution phases were initiated by a SOLVE THIS PART statement, the system is ready to accept modifications. The MODIFICATION statement is first examined to determine the initial action to be performed. If it is a MODIFICATION OF LAST PART, essentially no action is required, since all the input data associated with the previous part of the problem are available. If it is a MODIFICATION OF FIRST PART, the input data for the initial problem segment, which have automatically been saved, are restored to their original form.

After the initial action, the input continues, controlled by the CHANGES, ADDITIONS, or DELETIONS statements. Each of the latter statements sets a switch, which remains active until changed by another statement of this type. This switch controls the input processing, that is, to overwrite a previous piece of information, or add a new one, or delete an old one. Thus, when a new SOLVE or SOLVE THIS PART is encountered, the internal storage contains the description of a new problem, with the modified information superimposed on, or merged with, the previous data. The editing and checking functions of the compilation phase are then applied

---

[1] S. J. Fenves and F. H. Branin, Jr., "Network - Topological Formulation of Structural Analysis," ASCE Proceedings, 89, ST4 (August, 1963).

in exactly the same fashion as for the initial problem.

During the input of modifications, a tally is kept of the kinds of modifications specified, that is, whether they pertain to the structure, to the loads, or to the output alone. From this tally, the system can determine where to re-enter the execution phase. For example, if the structure remains unchanged, and only the loads have been modified, execution can immediately resume at step 3 described in the previous section. If, for some reason, only additional output is required, the system immediately jumps to step 5 (the SELECTIVE OUTPUT statement is handled as a special case of such a modification and does not require a MODIFICATION statement).

Since each modification is assembled with previous data and handled from then on as a new problem, an unlimited number of modifications are permitted, subject only to limitations of available computer time.

The process of alternating input and execution continues until a FINISH statement is encountered, which terminates a particular problem. The system is then ready to accept the next problem in the batch until the input tape is exhausted.

Appendix A

SUMMARY OF METHOD OF LINEAR STRUCTURAL ANALYSIS

The stiffness method of linear structural analysis used in STRESS solves a set of simultaneous equations relating the unknown joint displacement components to the known unbalanced force components at each joint.

For each member, a force-displacement stiffness matrix relating the member end forces and member end displacements is needed. The coefficient matrix of the joint displacements in the simultaneous equations is the joint stiffness matrix of the entire structure. This matrix is developed by adding for each joint the stiffness matrices of all the members incident upon that joint. The column vector of known unbalanced joint forces is found by adding the member fixed end forces to the joint loads. After this has been accomplished, the unknown joint displacements are found.

STRESS uses two different coordinate systems to minimize input computational error; one for joint quantities in global coordinates, the other for member quantities in local member coordinates. Both are orthogonal. The latter system is dependent upon the direction of the centroidal axis of the member and the orientation of the principal planes.

Calling JF the number of degrees of freedom of the joints in a particular structure, the transformation from one coordinate system to the other is accomplished by a JF $\times$ JF rotation matrix R, containing four submatrices:

$$R = \begin{bmatrix} \lambda & O \\ O & \lambda' \end{bmatrix}$$

The elements of λ are the direction cosines of one system with respect to the other. Since R is orthogonal, $R^{-1} = R^t$.

Consider a directed member AB, where A is the plus end and B the minus end; global coordinate systems are used at A and B as shown in Figure A.1. The stiffness matrix of member AB is a 2JF $\times$ 2JF matrix K^{AB} containing four JF $\times$ JF submatrices.

$$K^{AB} = \begin{bmatrix} K_{AA} & K_{AB} \\ K_{BA} & K_{BB} \end{bmatrix}$$

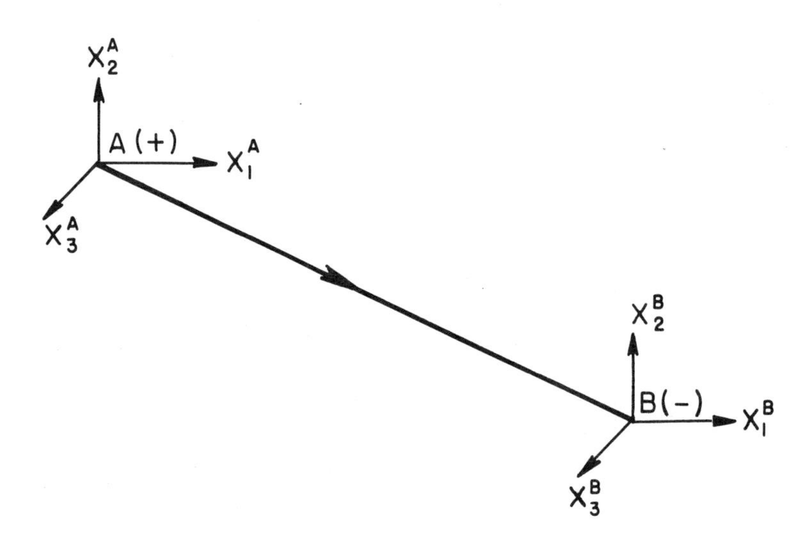

Figure A.1. Global coordinates for a member.

The elements in the first JF rows of K^{AB} are the force resultants at A and B arising from unit displacements successively applied at A with B fixed. During each application of a unit displacement, all other displacements are restrained. Similarly, the second JF rows of K^{AB} are the force resultants at B and A arising from unit displacements successively applied at B with A fixed.

It can be shown, however, that if K_{BB} is known, the other three submatrices can be obtained by an appropriate multiplication with a force translation matrix from end B to end A. This matrix T^{BA} is of order JF; its diagonal elements are unity, and certain off-diagonal elements contain the member projections in global coordinates. Now if K_{BB} is known in global coordinates,

$$K_{AA} = T^{BA} K_{BB} (T^{BA})^T$$

$$K_{AB} = T^{BA} K_{BB}$$

$$K_{BA} = K_{BB} (T^{BA})^T$$

The STRESS program computes K_{BB} for each member in local coordinates, rotates it into global coordinates, and performs the appropriate matrix multiplication to obtain the entire $2JF \times 2JF$ member stiffness matrix.

At this point, the member stiffness matrices of all members with force releases at either or both ends are modified, and the joint stiffness matrix is formed. Since the member stiffness matrices are in global coordinates, they may be added directly. For each member with the plus (A) end incident upon joint J, K_{AA} is added to submatrix location J, J (row and column position) of the joint stiffness matrix. Similarly, all members with their minus (B) ends incident upon joint J add K_{BB} to location J, J.

A member which connects joints J and K will add its cross stiffness matrices K_{AB} and K_{BA} to location J, K and K, J, respectively. It will be noted that the joint stiffness matrix is symmetrical as required by Maxwell's reciprocity theorem, as are the member stiffness matrices.

The presence of joint releases requires the modification of certain elements of the joint stiffness matrix. After this is performed, the joint displacements are determined by solving the equations. The solution of the equations in STRESS is by an "exact" method (successive elimination of unknowns) rather than an iterative or approximate procedure. With the joint displacements known, induced member distortions and end forces can be found by back-substitution. These induced quantities are added to the applied distortions and fixed-end forces computed with the joints locked to produce the final solution. Reactions may also be solved for once the joint displacements are known.

Appendix B

IBM 1620 STRESS LANGUAGE

A version of the STRESS programming system has been developed fo the IBM 1620. Due to the hardware limitations, this version is a subse of the original STRESS system, and has specifically been developed for use in education as a teaching aid rather than for general use in engineering practice. The restrictions inherent in the IBM 1620 version of STRESS as well as certain minor deviations in the input language from original version for the IBM 709/7090/7094 series are discussed below

Knowledge of the STRESS input language as explained in the first part of this manual is assumed, and some familiarity with the console operations of the IBM 1620 is helpful.

B.1. Machine Configuration

The IBM 1620 STRESS system was developed for a machine configuration with a core storage capacity of 40,000 digits and one disk drive. T configuration was selected because it is the smallest possible for a usa version of STRESS and is typical of many existing installations. In addi tion, indirect addressing, automatic division, the Move Flag instructio and floating point hardware are needed for immediate operation of the system. An installation must have all of the latter three special featur or some reprogramming will be required.

B.2. Restrictions

In contrast to the STRESS system for the IBM 7000 series, certain fi size limitations exist for the 1620 version. Such restrictions would not have been absolutely necessary, but they seem to represent a reasonab compromise between a very slow high capacity and a fast low capacity system. It is felt that such limitations are acceptable in view of the pu pose for which this system is intended. These limitations can be chang by the user in various ways. Specifically the fixed limitations on the si parameters can be changed with minor program adjustments if a 60,000 git core machine is available. The following is a list of restrictions isting for the present IBM 1620 version of STRESS:

B.2.1. Restrictions on size:

1. Maximum number of joints: NJ = 99
   (for any structure type)
2. Maximum number of members: NB = 200
   (for any structure type)
3. Maximum number of joint release components:
   NJR = 20
4. Maximum number of loading conditions in one
   loading combination: NP = 6

B.2.2. Restrictions on Capabilities and Input Language

1. No modifications can be specified, therefore no modification statements are available.
2. The PRINT DATA statement is not available.
3. Only the tabular form of data input is available for member and joint data.
4. Only PRISMATIC, STIFFNESS GIVEN, and FLEXIBILITY GIVEN are allowed for member property types. VARIABLE is not an admissible type.
5. In the JOINT RELEASE statements, no θ-angles can be specified. (They are assumed zero.) This means that the joint release coordinate system must be parallel to the joint coordinate system.
6. After the first LOADING header has been specified only loading data, loading headers, and output requests can be given until the SOLVE statement appears. All information not pertaining to loadings must be given before the first LOADING header.
7. Member distortions are not outputted.

.3. Language conventions

The STRESS input language of the IBM 1620 version differs from the M 7000 version in the following ways:

1. In the 1620 version of STRESS six or more consecutive blanks do not end the reading of an input statement. Instead, the reading is continued until the beginning of the next input statement is encountered. As many blanks and continuation cards as desired may be used in one statement to make the input more readable.
   Example:

```
    MEMBER        PROPERTIES        STIFFNESS        GIVEN
       6             6.3               0.0             0.6
       $             0.0              26.5           -16.4
       $             0.6             -16.4            18.0
```

2. The format of the COMBINATION loading header card has been changed. In the 1620 STRESS it is not necessary to have the word COMBINATION as the second word of the loading header for a combination loading. The word COMBINE is still needed on the data card following the header.
   Example:

LOADING	DEAD	PLUS	HALF	LIV
COMBINE	1	1.0	3	0.

3. The rules concerning specifications of output have been modified: if a general TABULATE statement is given before any loading header, and another TABULATE statement is given after the kth LOADING header, the output specification of this TABULATE command within the kth loading condition supersedes the general outpu command. The output specifications within the loading condition d not add to those of the general command.

4. The METHOD card may be omitted from the input program. The STIFFNESS method is the only method available.

B.4 Console operation

It is possible for the user of IBM 1620 STRESS to perform certain on-line debugging functions from the console typewriter. This on-line debugging option is selected by setting SENSE SWITCH 1 in the ON position at the beginning of a STRESS run. In this mode STRESS will echo-print on the typewriter any unacceptable statements that are encountered, together with a corresponding error code. A list of error codes follows. The typewriter then instructs the user "TYPE THE CORRECTED CARD OR RELEASE-START TO IGNORE." The user can then type the correct statement on the typewriter and press the Release-Start to resume processing; or he can press the R/S key directly if he wants to ignore the error. The user will be able to correct only errors of commission on line Errors of omission (such as leaving out the NUMBER OF JOINTS statement) are only detected after the SOLVE card is read, and are treated as fatal. After the first fatal error is met, the input data will be processed as though SENSE SWITCH 1 were off. In this case no on-line interaction is possible.

ERROR CODE	ERROR MESSAGE
1	FIXED POINT NUMBER IN WRONG PLACE
2	FLOATING POINT NUMBER IN WRONG PLACE
3	LOADING NUMBER GREATER THAN NUMBER OF LOADINGS SPECIFIED
4	NUMBER OF LOADINGS NOT SPECIFIED
5	END OF DATA FIELD BEFORE SUFFICIENT DATA GIVEN

ERROR CODE	ERROR MESSAGE
6	STATEMENT UNACCEPTABLE
7	(USED INTERNALLY BY STRESS)
8	THIS CARD IGNORED
9	ONLY LOADING DATA PERMITTED AFTER THE FIRST LOADING HEADER
10	MEMBER NUMBER GREATER THAN THE NUMBER OF MEMBERS SPECIFIED
11	JOINT NUMBER GREATER THAN THE NUMBER OF JOINTS SPECIFIED
12	ONLY SIX COMBINATIONS ARE ALLOWED IN ONE LOADING CONDITION
13	LOADING DATA INCORRECT
14	EXECUTION DELETED, FOLLOWING STATEMENTS SCANNED ONLY
15	NUMBER OF SUPPORTS NOT SPECIFIED
16	NUMBER OF JOINTS NOT SPECIFIED
17	NUMBER OF MEMBERS NOT SPECIFIED
18	JOINT INCORRECTLY SPECIFIED
19	STRUCTURE TYPE NOT GIVEN
20	REDUNDANT STRUCTURAL DATA
21	STRUCTURE TYPE AND RELEASE DIRECTION INCOMPATIBLE
22	ONLY MEMBER END LOADS ALLOWED ON MEMBER WHOSE STIFFNESS OR FLEXIBILITY HAS BEEN GIVEN
23-30	(NOT USED AT PRESENT)
31	NUMBER OF SUPPORTS NOT EQUAL TO THE NUMBER SPECIFIED
32	NUMBER OF JOINTS NOT EQUAL TO THE NUMBER SPECIFIED
33	NUMBER OF MEMBER PROPERTIES STATEMENTS NOT EQUAL TO THE NUMBER SPECIFIED
34	NUMBER OF MEMBER INCIDENCE STATEMENTS NOT EQUAL TO THE NUMBER SPECIFIED
35	NUMBER OF LOADINGS NOT EQUAL TO THE NUMBER SPECIFIED
36	STRUCTURE INCOMPLETELY SPECIFIED